The Community Life of Older People in Ireland

The Community Life of Older People in Ireland

Carmel Gallagher

PETER LANG
Oxford · Bern · Berlin · Bruxelles · Frankfurt am Main · New York · Wien

Bibliographic information published by Die Deutsche Bibliothek
Die Deutsche Bibliothek lists this publication in the Deutsche Nationalbibliografie; detailed bibliographic data is available on the Internet at <http://dnb.ddb.de>.

British Library and Library of Congress Cataloguing-in-Publication Data:
A catalogue record for this book is available from The British Library, Great Britain, and from The Library of Congress, USA

ISBN 978-3-03911-386-6

Cover images: Clareville Day Care Centre, Glasnevin, Dublin and location in County Donegal. Photo © Carmel Gallagher.
Cover images do not represent persons or places studied within the book.

Hochfeldstrasse 32, Postfach 746, CH-3000 Bern 9, Switzerland
info@peterlang.com, www.peterlang.com, www.peterlang.net

Printed in Germany

In memory of Dympna McGuire,
a good friend to many and a great
resource in her community

Acknowledgements

I wish to express my heartfelt thanks and indebtedness to Dr Ricca Edmondson, National University of Ireland, Galway who supervised the thesis on which this book is based. The enormous commitment of her time, advice and wisdom helped me to complete this work.

This study would not have been possible without the input of many people. I wish to sincerely thank all those older adults from Rathmore and Rathbeg who participated in the study. It was asking a lot to allow strangers in to your home to engage in discussion about important areas of your life and your locality. Your willingness to give your time, to contribute your experiences and your openness to exploring the ideas I was interested in are the foundations of this study.

I would also like to express my thanks to all those individuals from voluntary and community groups who gave me valuable insights into the everyday lives of older people. They include members of church groups, social clubs, voluntary groups, special interest groups, community development groups, youth clubs, staff and clients of day and residential care centres, church leaders and library staff.

I wish to convey my appreciation to the many officials and front line staff from health, housing and social care services in both Rathmore and Rathbeg who gave me their time, told me about their work, patiently answered my questions and engaged in discussion on issues that were emerging in the course of the study.

I wish to express my deep gratitude to Dr Teresa Brannick, University College Dublin for her critical advice and help with the quantitative analysis and for her personal support. Acknowledgement is also due to the Irish Research Council for the Humanities and the Social Sciences who awarded me a scholarship in 2002 to pursue this study.

I would like to express my appreciation to the Dublin Institute of Technology students who helped with the interviewing – Caitriona, David, Fiona, Katie, Linda, Roise and Tracey. The quality of their work was critical to the overall integrity of the study and they approached the task with great energy and empathy.

I wish to express my gratitude to my colleagues in the School of Social Sciences and Law, Dublin Institute of Technology, for their continuing support and friendship. I also wish to acknowledge the considerable support provided by the DIT which enabled me to carry out the study and to publish it. The support and library staff in DIT Mountjoy Square were always most helpful. I would like to acknowledge the help that Mr Ciaran Connolly, DIT provided with the SPSS analysis.

I would like to acknowledge the financial assistance which the Health Service Executive provided towards the publication of this book through Ms Edith Mullin, Service Manager for Older People Services, Donegal.

To my family and friends, I would like to say thank you all for your great encouragement. My deepest personal gratitude must go to my husband Noel and my mother Agnes for their unstinting support and feedback throughout.

Contents

Foreword

I would like to commend Carmel Gallagher and all associated with this fine publication – *The Community Life of Older People in Ireland*. A five year research project has resulted in an extensive and insightful study on the participation of older people within our society.

This study will enable policy and decision makers to better understand what it is that makes for a good life in older age and identify important issues for policy enhancement. It will also facilitate them in promoting policies and programmes that are based on a holistic view of older people in their family and community settings.

Working in both a rural and urban setting, the study shows the complexity which exists in modern Ireland in relation to the role that older people play and the challenges involved in integrating older people fully in our society. As a society, we are obliged to ensure that the older generations among us continue to play a full and active role within our respective communities.

Central to Carmel's thorough analysis is the cataloguing and recognition of the immense contribution that older people make to the lives of others. Although very often overlooked, the presence and work of older people is vital in our communities in a myriad of ways.

In addition to setting the parameters in which older people already make a valuable contribution, the study sets out the importance of local groups of different kinds in creating connectedness and building up a strong sense of community which many believe has been diminished in Ireland in recent years.

The study demonstrates the tremendous value of clubs and groups of different types in providing valuable social outlets in Irish communities. Such local groups provide a social network that strengthens the fabric of our society

This book alerts us to the importance of not taking for granted the value of volunteerism to our society. In line with the ongoing work of the Taskforce on Active Citizenship, this study identifies problems and solutions in enabling a strong sense of Irish citizenship as we progress into the 21st century. This study correctly identifies older people as having a central role in this process.

Along with the undoubted academic and analytical strengths which Carmel Gallagher brings to her portrayal of the role of older people in our society, this book gains an added depth with its basis in the lived experiences of older people in urban Dublin and rural Donegal.

Bertie Ahern, T.D.
Taoiseach

March 2008

Chapter One
Living in Rathmore and Rathbeg: An Introduction

Introduction

This book is an investigation of the lives of older people in Irish society. Based on extensive ethnographic and survey data, the book provides a sociological account of how older people live their lives and make sense of their lives at the start of the twenty-first century. It offers a descriptive account of the everyday lives of older people in different situations and in different settings in both urban and rural locations in Ireland. The book critically examines contemporary ideas about ageing, and in the light of the lived experiences of older people themselves, it provides insights into what a good life can be.

The study revealed that older people in Rathmore and Rathbeg are involved in a dense network of interaction with family, kin, friends and neighbours. Through formal and informal activities involving social and leisure activities, volunteering, informal helping, religious practices and informal sociability older people co-create positive forms of sociality. These activities and interactions are strongly associated with satisfaction, meaning and purpose in life. Face-to-face interactions within localities help to create enclaves of sociality where friendship is fostered.

While the study found high levels of connectedness there were older people who experienced disconnectedness. Some older people had accumulated emotional and material disadvantages owing to poor life chances. Difficulties that can contribute to social isolation and low morale include declining health, reduced mobility, death of spouse or friends, enduring or increasing caring responsibilities, insufficient health and social care supports and reduced income. While sociability and communal supports were evident throughout the study areas, there were areas where a combination of higher levels of stress (due to problems in families), poorer communal provision and environmental factors (such as vandalism) had contributed to lower levels of satisfaction and engagement with communal life. The study investigated how both connectedness and disconnectedness are experienced by older people, the conditions and processes involved in different forms of sociality and the contribution that older people themselves make to creating positive forms of sociality. Thus key elements of communityness are identified that are related to a good life for older people in Ireland at the start of the twenty-first century.

Aims and settings

The book provides both empirical and theoretical understandings of the nature of ageing and community in contemporary Irish society. It aims to provide a holistic account of older people's lives – their significant relationships, commitments and social activities. The book is based on a research study carried out between 2000 and 2005 titled 'The Participation of Older People in their Communities'.[1] It examines the social and community participation of older people[2] in two geographical communities in Ireland. Rathmore, an urban area in Dublin, was the primary research site, and Rathbeg, a rural area in the north-west of Ireland was the secondary research site.[3] The primary interest of the study was the participation of older people in their communities – their interactions and social activities and the extent to which these contribute to satisfaction and meaning in later life. The focus of the investigation is on the social and communal activities of older adults including leisure interests, involvement in clubs, religious practice, voluntary work, relationships with kin, friends and neigh-

1 'The Participation of Older People in their Communities', unpublished PhD study, National University of Ireland, Galway, 2005.

2 A cut-off age of 65 and over was chosen for the purposes of this study since this is the normal retirement age in Ireland.

3 Rathmore and Rathbeg are fictitious names.

bours, help given and help received, use of social services and informal interactions in neighbourhoods and other communal settings. Key themes explored in the study are the importance of friendship, the contribution of older people to the lives of others and people's own understandings of what is a good life.

People grow older in a particular place and we cannot explore either what a full life might mean or processes related to ageing unless we seek to understand the context and setting in which people live their daily lives. Two contrasting geographical locations were chosen to investigate ageing processes in a community context. Rathmore is a large suburb in Dublin, while Rathbeg is a small village in County Donegal. The study sought to examine both individual patterns of social and communal activities among a diverse group of older adults and to map the landscape of social groupings in Rathmore and Rathbeg. A multi-method research strategy was employed, including the use of ethnography, survey and interviews. Different approaches were used at different stages of the research that aimed to provide understanding of older people's lives in the context of their relationships and commitments. By combining ethnographic and statistical analyses we were able to obtain a detailed disaggregated picture of how older people live in their communities, to explain patterns of connectedness and disconnectedness among a diverse population of older adults, and to contribute to theoretical understandings of processes related to ageing and community. We were thus able to map the significant social groupings in Rathmore and Rathbeg, to delineate the relationships of sociability and support involving family, kin, friends and neighbours, and to provide rich description of activities and relationships that were meaningful in the lives of older people.

The lives of older people have become subject to greater political and research scrutiny over recent decades in the industrialised West. This is partly the result of socio-demographic forces that have led to a 'greying' of the population and the policy implications that arise from this. It also results from critical stances towards aspects of contemporary culture and a questioning of the assumptions about age which are embedded in our institutions and our practices. Of the different theoretical and empirical approaches in sociology that seek to illuminate the experiences of older people in contemporary society, many approaches tend to depict lives in very individualistic terms. Others emphasise the influence of cohort and class on experiences of ageing. However, it is the interpersonal aspect of ageing lives and the extent to which meaning and satisfaction in later life are related to communal life which are the focus of this book.

Ireland is often referred to as a post-modern or post-Christian society that has undergone transformation from a traditional society to a modern, industrial and secular one. Such changes are evident in rapid economic growth, the ending of mass voluntary emigration and the arrival in Ireland since the mid-1990s of significant numbers of immigrants. However, the reality of social change is more multi-layered, with elements of past and present co-existing (MacEnri, 2006). This study provides insights into the nature of community in a rapidly changing society and into older people's contributions to their communities. It points to realms of activity that offer genuine meaning and value in older people's lives and explores the meaning of connectedness between them. Thus it portrays a relational landscape which can seem in danger of being ignored in Irish and other contemporary societies. At the start of the twenty-first century this study is a useful baseline against which future changes can be examined.

Older people and community

Our interest in the communal aspects of older people's lives stems from a desire to ask questions about the actuality of older people's lives and the possibilities for older adults to live full lives in respect of both their own personal development and in the ways they might enhance the lives of other people. The types of questions that arise include: What are the lives of older people like in contemporary Irish society? What kinds of interdependencies exist in neighbourhoods and communities? What is the nature of connectedness involving family, kin, friends and neighbours? To what extent are older people in different localities living satisfying lives? How do older people themselves define a 'good life'? Such questions required us to examine both the nature of ageing and the nature of community. While a broadly humanistic approach is used in the study, it is important to point out that at the outset it was not at all clear to what extent and in what ways communality might be important in understanding the lived experiences of older people. Recognising that ageing is, at least in part, a socially constructed phenomenon, was it possible to obtain insight into people's lives in a way that both retained the individual voices of older people in different situations and with different accumulated life experiences, while at the same time contributing to theoretical understanding of the nature of sociality and connectedness in contemporary society?

It was hardly surprising that different experiences and perspectives were found among the participants of the study in relation to questions of social and communal participation. This is illustrated by the example of contrasting pictures presented by two study participants in Rathmore of the social and communal engagement of other older people in their immediate locality. One person painted a picture of contented lives among the fellow residents of a sheltered housing complex:

> They all seem to be happy doing their own thing. They keep the place tidy and go where they want … There are social activities at [name of day centre] if they want. At night time you don't want to go out. I talk to them all. They all seem to be happy, well dressed and keep themselves well (Interview with study participant, Rathmore, July 02).

A less positive picture of older people's lives in Rathmore was presented by another informant who was socially active herself, but believed that there was a lot of isolation and boredom in the lives of older people:

> There's a huge amount of idleness and hanging around among older people … an attitude of 'Ah sure, I'm past it … what would you want to get that done for?' People are living more separately … there's isolation with age … no structure to your day (Interview with study participant, Rathmore, April 02).

With contrasting perspectives such as these, based on different life experiences, life circumstances and experiences of sociability, there were clear challenges in attempting to develop sensitive and nuanced concepts related to communality and in proposing explanatory models of connectedness that described and explained the experiences of a diverse group of older people who participated in their communities in different ways.

The evolution of the concept of community in the social sciences reflects historical changes in social relations, different moral and value stances and dilemmas in relation to the human condition, such as the need for both constancy and support of a community as well as the desire for personal development and social change. The Donegal poet, Cathal O'Searcaigh, writing about his native place, weaves together the themes of interconnectedness, communal constraints and openness to others found in a small-scale rural community. In the tight-knit 'pobal'[4] of Caiseal na gCorr there is no sharp differentiation between personal choice and relationships and public interactions. O'Searcaigh sees this type of community as characterised by openness and kindness among people:

> It was of course a valley of squinting windows. An obsessive inquisitiveness about each other made people peep and snoop and eavesdrop. This sort of community was, and still is, like a spider's web. Everything and everybody is intimately interconnected. If you seriously transgress, let's say, by pulling a thread here, the jerk of that misdeed, the knock of that offence, will reverberate throughout the entire community network … the *pobal* sets standards of conduct, models of behaviour, certain yardsticks that outline acceptable actions and attitudes. If you are born and reared here, you understand all of this instinctively. It's in the air. It gets transmitted to you anyway, by some subtle social osmosis. Despite the fact that we have become more affluent in recent years, there is still a strong sense of community here in the glen. I have always noticed that when people become well off they demand the privileges of privacy. They become withdrawn. They migrate to their own private inner spaces. This can endanger the sort of open, fluid, sinuous style of interaction that we have practised in the Glen for many a generation. May a door of kindness be always open in our hearts (Voster and O'Searcaigh, 2003: 53).

The theme of kindness and community was strongly expressed in a speech by President Mary McAleese, Uachtarán na h-Éireann to the

4 People or community.

Conference of Religious in Ireland (Address by the President, June, 2005). She extolled the work of religious and voluntary groups as an expression of connectedness and what she termed 'the human infrastructure we offer each other through friendship and community solidarity.' Her view of community is 'a two-way traffic of decency and goodness.' She suggested that as a nation 'we are good at connectedness, that 'teangmhail', that human spark which spreads from one of us to the whole community.' However, echoing concerns by writers such as Robert Putnam about decline in 'social capital', she also suggested that community needs to be recreated in each generation and that 'it would be sad if in winning our prosperity we lost our time for one another.'

Yet, to what extent are the above visions of caring and cohesive communities expressive of the *idea* of community rather than descriptive of the *reality* of modern, urban communities, especially in the face of rapid social and economic changes? The so-called communitarian crisis has been articulated by writers such as Putnam, Etzioni and Fukuyama who have expressed concern that the old bulwarks of community – class, kinship and neighbourhood – are disappearing. The idea of older people as the 'roots' of communities – key elements of solidarity and continuity in a changing world – has been challenged by 'modern' retirees who enjoy the possibilities of choosing an alternative place and style of retirement (Gilleard and Higgs, 2005). While we might assume that communities of propinquity are more salient in the lives of older people than younger adults, it is far from clear how communality is influenced by place as opposed to common interests or common identity.

Concepts and themes

Theoretical concepts that seemed useful in understanding the meaning of lived experiences included the concept of the life course that views the latter stages of the life course as a dynamic process linked both biographically and historically with earlier life stages (Hareven, 1996). Such perspectives as well as those from political economy (Estes, 1979; Walker, 1981; Townsend, 1981; Phillipson, 1982, 1998; Estes *et al.*, 2003) help us to understand the socio-cultural and the socio-economic factors that influence the possibility to become one kind of person in old age more easily than another. Humanistic perspectives give insight into what might constitute a good life in old age and to express the potential for human beings to continue to develop and to contribute positively in their older years (Erikson, 1980, Erikson *et al.*, 1986; Moody, 1988, 1998).

Research interest in the lives of older people in Ireland has been growing. Much recent research has been of an applied nature focusing on aspects of the quality of life including health, social care supports and well-being of older people both in their homes and those living in long-stay care (Garavan *et al.*, 2001, Murphy, 2006, Age and Opportunity, 2003). However, a criticism of empirical approaches to measuring adjustment and well-being is that they do not take account of the social context of lives and develop rather static indicators of processes that are complex and to portray the older person as a rather passive possessor of certain attributes (or recipient of social goods). What has received less attention is any investigation of the nature of connectedness and the intersubjective experiences of older people in a particular place. This calls for a more relational level of analysis and is often a missing ingredient in accounts of individual adjustment to old age or of the broad social forces that provide the context for ageing.

These theoretical approaches guided the design of the study. However, the theoretical perspectives that directed us what to look for and how to interpret what we saw evolved in a dialectical relationship between analytical concepts in the sociological literature and the views and ideas actually used by people who lived in the study areas. Therefore, the meaning of community itself became the key analytical idea throughout the study, aspects of which were subject to critical analysis and examined in relation to how they appeared to resonate with the empirical data. Drawing on classical accounts of community by writers such as Tonnies and more contemporary accounts of the nature of sociality in contemporary capitalist, globalised societies, we examine concepts such as kinship, social network and neighbouring. The concept of friendship emerged as a key analytical idea that helped to explain the significance of relations in socio-spatial settings and the types of experiences that give meaning to older people's lives. A particular conceptualisation of friendship is proposed arising out of the study that challenges some contemporary accounts of friendship in modern capitalist societies.

Methods of inquiry

Our desire to investigate the interdependencies and reciprocities that existed within two localities and to both describe and understand the different types of connectedness and disconnectedness among different groupings of older people required a methodological approach that was drawn from a number of different traditions and that developed throughout the duration of the study. There are particular methodological challenges involved in research on older people when trying to understand processes of life-course construction. Furthermore, no piece of research can claim to fully understand all aspects of a phenomenon and researchers can only do their best with the theoretical and methodological tools available and within the resource constraints of a particular study.

Nonetheless, it seems doubtful that a researcher could come close to understanding the meaning of social and cultural practices of older people or older people's responses to questions about their daily lives without taking into account the following:

1. Biological and historical circumstances of earlier stages of the life course,
2. A referent community, in some sense of the term to be investigated further,
3. Communicative forms used by older people themselves, and
4. Institutional practices as sites of power and influence.

Thus we see value in theoretical approaches that adopt a life course perspective, that view older adults as creators of their own lives in interaction with others with whom they share life space, but at the same time recognise the societal constraints and cultural assumptions that shape possibilities for various types of action and participation. The diversity of the population making up the 'third age' and the 'fourth age'[5] and the nature of the questions to be asked suggested that the study should seek both breadth and depth. For this reason a multi-method research strategy was employed involving both qualitative and quantitative methods of inquiry. In this way it was hoped to capture older people's own experiences in Rathmore and Rathbeg while trying to obtain broadly generalisable knowledge about patterns of participation across different groupings in these localities.

Recognising the importance of the links between emergent theoretical positions and the way many of the key variables used in research on ageing are defined and measured, the value of using a combination of methodological approaches as well as analytical approaches derived from grounded theory was suggested. Our approach to developing methodology was, therefore, to see it evolving as part of the process of research and part of the research problem itself i.e. how to understand the experiences and meaning of ageing for a diverse group of older adults. Thus, the research process, derived from grounded theory, involved continuous formulation and reformulation of concepts in an ongoing dialectical relationship between the ideas and views of older people and the analytical concepts. This approach underlined the premise that both researcher and researched were to some extent involved in the creation of knowledge.

5 The Third Age has been suggested as marking the period when an individual has retired from formal work but is physically and mentally active and able to engage in active leisure. The Fourth Age is a stage of final illness and/or dependency leading to death.

Methods used

The principal methods used in the study were:

1. An interview-based survey was conducted with one hundred and sixty five older people in Rathmore. The interviewers called to houses on a random basis and steps were taken to ensure that a quota of older people from all the principal sociological categories was obtained (for example, age, class, gender, neighbourhood, living alone or co-resident and type of accommodation). A small comparative survey was carried out in Rathbeg involving interviews with eight older people.
2. Ethnographic work was carried out in a range of settings in Rathmore and Rathbeg including Senior Citizens' clubs, Active Retirement Associations, day centres, voluntary groups, special interest groups and communal settings such as pubs, churches and social centres. This involved observations and informal interviews with participants, organisers and staff. Observations became more focused as theoretical themes emerged.
3. Interviews were carried out with key informants including social services and health care staff, volunteers, members of clubs, religious leaders and community activists.

Choosing the research sites

Rathmore

Rathmore in Dublin was the primary site chosen for the study and Rathbeg in County Donegal was the secondary site. Rathmore was chosen as the primary study area for a number of reasons but the following four considerations were paramount:

1. It contained large numbers of older adults,
2. It was relatively socially mixed,
3. The researcher had some prior knowledge of the area, and
4. Its boundaries correspond with official administrative boundaries for which demographic data is available.

Rathmore is not a natural community in a geographic sense nor would people living within its boundaries regard those boundaries as significant. We sought a locality which had a higher than average proportion of older people, relatively long-established households, a social class mix, some sheltered housing for older adults and a range of services and activities that are used by older people such as clubs, day centres, pubs, churches and leisure amenities. The study area chosen, Rathmore, about four miles north-east from Dublin's city centre, is composed of four adjoining District Electoral Divisions. The District Electoral Division is the smallest administrative unit for which population statistics are available, therefore, the search for a suitable site started off by examining Small Area Population statistics (SAPS) from the most recent census in 1996.

Rathmore's population size (12,700 persons), suburban features (large housing estates with amenities of public transport, public parks and services) and area (about 4 square miles) made it both suitable and practical for carrying out ethnographic and survey type research. Of the total population, 2,483 or 19.55 per cent were aged 65 or older. (SAPS, Census 1996). In the 1996 census 11.5 per cent of the national population was over 65 years. While there is no such thing as a typical community, it was not desirable that the study area should be identifiably untypical compared with other communities. We believed that the more heterogeneous the area in terms of different household compositions and age groups, different social classes and availability of the most common services, amenities and activities used by older adults the more possibilities there would be to explore all aspects of participation and to specify the conditions under which particular types of participation are likely or not likely to occur. The researcher used personal knowledge, professional contacts with social care and health care staff and visits to the area to ascertain that the area was indeed a suitable site to investigate processes related to ageing and community.

Sheltered housing

In addition to the general population of older adults living in ordinary housing it was decided to include sheltered housing complexes in Rathmore as a specific site of investigation. It is primarily older adults who live in these complexes since sheltered housing represents a particular public policy response to the accommodation and general social needs of some older people. There is also a sense in which a sheltered housing scheme may be regarded as a form of constructed community. For these reasons it seemed worthwhile to include residents of sheltered housing both as a site where important processes related to ageing and community could be explored and to investigate a particular aspect of public policy towards older adults.

Rathbeg

Rathbeg was chosen as the secondary research site because it provided a contrasting socio-geographical location to the main study area, the author already had relevant knowledge about the area and it was expected that access to key people and settings could be obtained. Rathbeg is situated in County Donegal, a county in the north-west corner of Ireland with a population of approximately 130,000. Rathbeg itself has a population of 812 of whom 118 or 14.5 per cent are aged 65 or over (1996 census). The environs of Rathbeg are rural, consisting of open countryside and other similar sized villages. However, the principal town with a population of 16,000 is only 20 kilometres away. Rathbeg is the birthplace of the author and she spends significant amounts of time there. Therefore, it seemed to provide an opportunity to build on her innate knowledge and connections there to complement the research work in Rathmore. For example opportunities to make observations abounded in the author's own social life when for example at church, in the pub, at funerals, visiting friends and neighbours, in shops, attending community events and while out walking. While extensive ethnographic work was carried out in Rathbeg and its environs, the quantitative work was much more limited. The quantitative study in Rathbeg was not designed to replicate the survey in Rathmore or to compare item by item with the results of the Rathmore survey but rather to provide some points of comparison in relation to broad differences and similarities in the lives of older people in an urban and a rural locality.

Stages of the study

Ethnographic work

Extensive ethnographic work was carried out in both areas involving observations in many diverse settings and interviews with older people and key informants. Initially the author chose to visit as many different types of settings as possible and later as the study became progressively focused the settings became more selective. Observations were made in day care centres, sheltered housing complexes, social clubs, active retirement groups, residential nursing homes, voluntary groups, churches, shopping centres, and other settings where older adults socialise. The author drew up lists of groups that were mainly associated with older adults such as Active Retirement groups and Senior Citizens' groups as well as groups that were not necessarily associated with one age group such as Ladies' clubs, community development groups and special interest groups. Useful sources of information were notice boards in churches, shops, social clubs and libraries, parish newsletters, parish directories, community newsletters, and brochures in social services centres. While it was not possible to make contact with all groups in Rathmore the author sought to make contact with at least one group from what appeared to be the main broad categories of social groups in which older people were involved.

Opportunities for participant observation were found at different times during the study. In Rathmore the author attended many different types of social gatherings over the course of the study. The following are examples of some of those activities:

- Helped with the preparation and serving of a meal in a sheltered housing complex.
- Attended three meetings of a church-based group to set up a visiting scheme to lonely older people.
- Participated informally in activities in day care centres, Active Retirement groups (bowling activity, art session, bingo, sing-songs etc), community festival events and social gathering of Ladies clubs.
- Attended meetings of Local Historical societies and community development groups.
- Visited local pubs in the company of an informant to observe and meet regular pub goers.

A parallel process of research activity was initiated involving interactions between the researcher and the 'researched' in Rathbeg. Similar methods to those used in Rathmore were employed to identify the significant social groupings in Rathbeg and its environs. The author visited social care settings and conducted interviews with key informants including social services personnel, community activists, voluntary workers and members of clubs. The author spent time in three different day care centres observing the daily routines, interviewing staff and volunteers, engaging in informal dialogue with service users, and taking part in some recreational activities.

Interviews were carried out with older people from different backgrounds that explored their concerns and the significant relationships and commitments in their lives. Interviews were also conducted with key informants who because of their work had knowledge of the lives of older people.[6] Furthermore, some of the informants were

6 Interviews were conducted with the following key informants: Health and social services personnel, Committee members of several voluntary groups, Local Parish priest, Lay Minister Church of Ireland, Community development worker,

themselves in the older age group. The author monitored local sources of news and events throughout the period of the study and sought out opportunities to inform herself of events and activities and to attend events relevant to the lives of older inhabitants. Events attended included a centenarian birthday celebration, an art exhibition and community festivals – all of which provided insight into communal life and aspects of ageing. Field notes were made after each visit and event. The author's mother, other older relatives and friends were also a great repository of knowledge and insight.

In the case of both Rathmore and Rathbeg, there was a concentrated period of immersion in the initial phase of the study, which gave the author an overview of day-to-day life, knowledge of the relevant institutional settings and of aspects of sociality that appeared important in the lives of older adults. However, interaction with older people and contact with social groups continued throughout the period of the study. Contacts at later stages were directed to specific groups and settings as themes became more progressively focused. These contacts served to obtain feedback from participants, to validate the initial conceptualisation and to give direction to subsequent work and analysis.

Survey

A large-scale survey was carried out in Rathmore and a small survey in Rathbeg. The survey was designed as a creative research instrument that would obtain both quantitative and qualitative data, in order to achieve both a broad overall descriptive understanding of community participation and also deeper understanding of the meanings of different forms of participation and experiences of community. The strategy adopted was to aim to construct a particular type of questionnaire and associated interviewing strategy that involved a dialogue between interviewee and interviewer. While in conventional survey research the interview is seen as a pipeline for the transmission of knowledge,

---

Female activist in older persons' community development project, Member of local special interest groups.

here the interview was designed to be more interactional and to produce more elaborate narratives of people's lives. The questionnaire contained open-ended questions and recorded the direct words used by the respondent. Specific steps taken to build rapport with the respondents included sharing information with them on the purposes of the survey, sensitivity in framing questions and undertakings to give feedback on the results of the survey.

Questionnaire

The main Interview Schedule contained 40 questions (See questionnaire at Appendix A). A variety of different types of questions were asked including fixed response questions, semi-structured and open-ended questions. The research themes that emerged from the initial stages of the research were used to formulate questions in the following areas:

- Functional interdependencies,
- Patterns of work, leisure and social activities,
- Experiences of retirement,
- Helping relationships in daily living,
- Meaning of community,
- Older adults as a resource,
- Civic engagement,
- Factors related to participation,
- Religion, spirituality and value frameworks,
- General experiences in relation to the role and social participation of older people.

Many drafts of the questionnaire were drawn up before it was ready for testing. Two pilot rounds of interviewing were carried out before the questionnaire was finalised.

A section was created in the questionnaire (Part B) to enable the interviewer to record any impressions they gained of the interviewee. This was recorded independently by the interviewer when the interview was completed. It was designed to record anything striking or especially interesting about the respondent which might help in interpreting their answers or which might suggest a theme which could be followed up later.

A Preamble was drawn up which explained the rationale for the survey and gave background details on the research and the researcher. It was used by the interviewers when they first called to the door in their introductory conversation and then given as a handout for the person to keep The Preamble explained the aims of the survey, the topics explored in the questionnaire and the potential benefits of the data acquired. It also indicated that a follow-up meeting would be held to disseminate and discuss the information obtained to which all respondents would be invited. The Preamble was considered an important part of the survey in two respects: firstly, it underlined the social responsibility of the researcher towards the survey participants and the community generally and secondly, it allowed some initiative to the respondent.

Two other short sections were included. Part C of the questionnaire was used to record details of the respondent's accommodation and general surroundings. Part D was used for a short summary of the interviewee, including age, sex and whether living alone or with others. This enabled the researcher to monitor the types of respondents who were being sampled to ensure representativeness.

Sample

The survey was used to locate older people who were living 'ordinary' lives and were neither clients of social services nor necessarily involved in the social groups we had mapped. In the absence of a suitable sampling frame, calling randomly at houses in Rathmore seemed the best way to obtain reliable knowledge about the everyday lives of a heterogeneous grouping of older adults. The sampling strategy adopted represented a mix between purposive sampling and random probability sampling. Thus a sample that was representative as far as possible of the principal subgroups of the general population of older people would facilitate comparison and some degree of generalisation to the wider population. Therefore the aim was to obtain a sample that would include older adults from all four DEDs,[7] men and women, people from different social classes, younger-old (65–74 years) and older-old (75 years and upwards), people living alone and people who co-resided, residents of sheltered housing and residents of ordinary housing. A sample size of 150 was thought to be sufficient to capture diversity and enable comparisons to be made between subgroups in a statistically meaningful way. Based on the Small Area Population statistics, we established quotas as a broad guide to check whether any group was significantly under or over represented. Over a three week period in summer 2002 the interviewers (including the author who conducted a sizeable number of the interviews) were deployed to call at houses (initially every 3rd or 4th house) and seek an interview with any household member who was aged 65 or over.

Sampling of the residents of sheltered housing was conducted in a different way. In consultation with the Community Liaison Officers with responsibility for the complexes, interviews were arranged in advance as some of the residents might be fearful and suspicious if an interviewer about whom they had no prior knowledge called to their home.[8]

Over a period of three weeks one hundred and sixty five interviews were obtained altogether. This was considered very satisfactory for purposes of generalisability and, insofar as the quota guidelines were concerned, the sample contained a good 'mix' of respondents. There were almost equal numbers of young elderly and older elderly. In relation to gender, 45 per cent of the sample was male while 55 per cent was female. Those living alone were over represented somewhat, however, this was largely accounted for by the inclusion of 18 residents of sheltered housing, all of whom lived alone. The sample was reasonably representative with regard to marital status and socio-economic diversity. While the largest proportion of interviews was

7 Rathmore consists of four District Electoral Divisions.

8 The author specified the number and 'type' of resident she wished to interview – 18 persons altogether, six residents in each of the three complexes. The Liaison Officers assisted with the selection from each complex of a younger (64–74) female and younger male, an older (75+) female and older male, and a male and female user of health and social services.

obtained in two of the DEDs, one of these DEDs contained more middle class neighbourhoods and the other more working class neighbourhoods, thus ensuring a social class mix of respondents.

Survey in Rathbeg

The quantitative study in Rathbeg was not designed to replicate the survey in Rathmore or to compare item by item with the results of the Rathmore survey but rather to provide some points of comparison in relation to broad differences and similarities in the lives of older people in an urban and a rural locality. Interviews were conducted with eight older residents of Rathbeg using the same questionnaire instrument. The sample was purposive and interviewees were selected on the basis of similar criteria used to obtain a diverse sample in Rathmore – gender, social class background, age and whether living alone or co-residing. The author conducted all the interviews herself in the respondents' own homes. Given the author's familiarity with Rathbeg all the interviewees were known to her. The researcher felt somewhat uncomfortable asking certain questions of people she knew such as those related to help received from and contact with family and relatives. She stressed in advance that the respondent should not feel obliged to answer any question that he or she was unhappy about. In the event no respondent declined to answer any question although one person was anxious for reassurances that what was said should in no way identify that respondent. Because of the small size of the sample and the possibility of being able to identify the respondents it was decided not to offer feedback. Furthermore, very scant details are given of the eight interviewees to protect their anonymity in a small locality.

Engaging with respondents and establishing trust

We consciously set out to use the relationship between the interviewer and interviewee to obtain reliable knowledge that would bring the interviewer into the interviewee's world. At the same time we were conscious of our ethical obligation to inform and to ensure insofar as we could that the experience would be enjoyable for both parties. We were also concerned that we should give something back to those who took part. The principal ways we sought to achieve these aims were:

- Training the interviewers,
- Drawing up a written protocol to ensure the interviews were conducted in a sensitive and honest way,
- Ensuring that the questions were appropriate and sensitively asked,
- Giving information prior to the commencement of the survey,
- Preparing a handout on the purposes of the survey, and
- Giving an undertaking to arrange a meeting after the survey to give feedback on the results.

The fieldwork experience was very positive in regard to responses to the requests for interview, the quality of the interviews obtained, the reported nature of the interactions between interviewer and interviewee and the hospitality that many of the interviewees displayed to the interviewers.

Living in Rathmore

History

Rathmore was originally a rural village with ancient origins. It is thought to have been an important settlement in Viking Ireland in the 11th century. The history of Rathmore has been recorded by the local Historical Society in a number of publications. The common name associated with Rathmore is derived from the Gaelic and is believed to relate to an Irish saint from the early Christian period in Ireland. In the late 18th and early 19th centuries Rathmore was a very rural area comprising large private estates, small farms and labourers' cottages. Many of the 'Big Houses' were occupied by families who were prominent in the social, economic and political life of the country. Many of these old houses are still there and are being used for example as a school, a hospital and a retirement home. The remaining 18th and 19th century houses, the railway station and Church of Ireland church which were built in the 19th century, and other remains of older buildings, create a village-like atmosphere in this part of Rathmore. Rathmore remained largely unchanged until the spread of Dublin city in the 1950s.

Rathmore today

The area of Rathmore is today dominated by suburban housing estates. New housing built in the 1950s and 1960s, both public and private, brought many young couples and families into the area. Many of the older adults have lived in the same house which they first occupied forty or fifty years previously. The vast majority of respondents in the survey have lived in Rathmore for at least 30 years and in almost all cases in the same house – over 80 per cent had lived in the locality for at least 30 years and 46 per cent for at least 40 years.

Rathmore can be described as a settled mixed social class area, comprising mainly lower-middle and upper-working class groups. There is no industry located there apart from service employment, however, there are small industrial estates in its environs. All the major occupational groupings are to be found there.

The shopping/commercial area still retains village-like features, including the remains of historical buildings and monuments, old houses and a railway station. While Rathmore is now entirely built up, a large mature public park with pleasant walkways and magnificent trees and shrubs adjoins one part of it and at the nearest point Dublin Bay and its sandy beaches is only a quarter of a mile away. Some streets adjacent to the oldest part of Rathmore or adjoining the large public park are surrounded by trees. The large local-authority-built estates are generally in open spaces with few trees.

Housing

Local authority housing estates were built in different phases approximately between 1954–1965. Private housing estates in Rathmore were developed mainly during the 1950s and 1960s. While the private houses tended to be built nearer the main roads and closer to the village centres, the local authority estates were generally built in the large green spaces in between. Practically all the local authority housing in Rathmore has been purchased under Tenant Purchase schemes and is now owner-occupied. There have been some new small private housing developments in more recent decades and a few apartment complexes built as remaining pieces of land have been developed. The generally well-maintained houses and gardens indicate a reasonable degree of material comfort and ability by residents of both public and private housing to manage the up-keep of their homes. The private houses tend to be bigger and to have more spacious gardens than the houses built by the local authorities.

Families who moved to Rathmore in the 1950s and 1960s as young adults in their 20s are now in their 60s and 70s and their children are well into their adulthood. Many respondents recalled moving in to their houses on marriage or when their children were very young. Many houses have gone up for sale in recent years, particularly in the areas with the oldest housing and which have higher proportions of older people. Priests in the area say they have a large number of funerals. With the current high property values prevailing in Ireland the price of both local authority and privately built houses has soared in recent years.

Sheltered housing

There are three sheltered housing complexes in Rathmore which were built by Dublin City Council between 1968 and 1977 with approximately 50 to 60 units in each. The units are allocated under a scheme of Priority for Letting.[9] The age limit for entitlement to sheltered housing used to be 60 years but this was recently reduced to 55, bringing a somewhat younger age group into the courts. The units in Rathmore are all bedsits (apart from a few double units for married couples) and the accommodation consists of a living room/bedroom, a small kitchenette and a bathroom. The units are very small, approximately 180 sq. ft., and the absence of a separate bedroom is a major drawback for many of the residents. There are Housing Liaison Officers (formerly known as Wardens) employed by Dublin City Council to provide what the Council literature terms a 'good neighbour' service to the residents.

Amenities and facilities

Rathmore has a large public park with walk-ways, football pitches, pitch-and-put and tennis courts. Rathmore is generally well served by public transport – there are scheduled routes served by Dublin Bus and two DART train stations (light electric train service). Other amenities in the area include schools, library, sports clubs and a swimming pool. There are a variety of clubs and formal groups in Rathmore, including social clubs, special interest groups and church-related groups. Clubs meet in parish and community halls. Groups that cater specifically for older people are Senior Citizens' groups and Active Retirement groups. Pubs are also an important venue for older people, especially older men. Older people can be observed in public places in Rathmore using public amenities, attending church services, on the buses, in local shops and post offices and participating in club activities. There are four Catholic churches in Rathmore serving a predominantly Catholic population. The Protestant community served by two churches represents a very small proportion of the population. In relation to social care type facilities Rathmore has one Day Care

9 Units are currently allocated from four lists: medical grounds, homeless status, buy backs (a scheme where the Council buy the applicant's house and allocate them a unit) and the ordinary waiting list (information at meeting with Council official on 21/10/03).

Centre operated by the Health Service Executive (at the time of the study this was the Health Board). There are two long-stay residential homes for older people.

Living in Rathbeg

History

Rathbeg is a small-scale community with a long history and long-established traditions. In Rathbeg, for example, some residents can trace their family connections in the locality back to the late 1700s. There are many examples of settlement of more ancient origins. A local historical society has collected and recorded local history of the twentieth century from photographic, oral and written sources. Preparations to commemorate a major national historical event that took place in Rathbeg were underway while the study was in progress. As in many rural areas in the West of Ireland, emigration of young people to urban parts of Ireland and abroad has been a feature of life in Rathbeg during the 20th century and before.

Rathbeg today

Rathbeg is a scenic seaside village with a farming hinterland whose permanent population increases during the summer months and at other holiday periods. The local community comprises people who have lived in Rathbeg all their lives, people from Rathbeg who worked abroad for part of their lives, typically in England or Scotland, people who grew up elsewhere but chose to live in Rathbeg, perhaps on retirement, people who grew up elsewhere but married and settled in Rathbeg. There is a large 'diaspora' of people born in Rathbeg and living in different parts of the world from the United States to New Zealand. They return regularly to visit their families. Some have bought second homes in Rathbeg and would regard themselves as belonging to the community by virtue of growing up there, their family and friendship ties and frequent visits. While Rathbeg may be thought of as more self-contained than Rathmore, it has experienced significant social change in both inward and outward movement of people.

There is no major industry in Rathbeg. However, farming is a long-established livelihood of many Rathbeg families. Work has also been available in construction due to tourism and increased second home ownership. The main occupational groupings can be identified in Rathbeg – professionals, proprietors – mainly of small businesses, farmers, white-collar workers who work in offices in nearby towns, manual workers in the fishing and building industry, seasonal hotel workers and casual workers on social employment schemes.

Housing

Most of the housing in Rathbeg is owner occupied, as is usual in the Irish countryside. However, there are two small public housing estates, one built in the 1950s and the second around the year 2000. A small number of houses for 'social housing' were built by a voluntary housing body in 2002. A number of public housing units and voluntary housing units have been specifically allocated for older people. In addition to permanent residences large numbers of holiday homes have been built in recent years. Many of these are second homes owned mainly by people from Northern Ireland who use them at weekends and at holiday times.

Amenities and facilities

Amenities include local walking pathways, safe swimming beaches, a small marina, a football pitch, a nine-hole golf links and a small resource centre. A limited public transport system operates. One bus travels to the principal town early in the morning and returns in the evening. There are four pubs and two hotels. At the time of the study there was no community or parish centre, however a community centre has recently been opened. The pubs are used for occasional social events such as table quizzes and fundraising coffee mornings organised by local groups. There are many social groups including church-based groups and groups involved in different aspects of community enterprise and development. There are two main Christian communities in Rathbeg, Catholic and Protestant, the Protestant community being the minority. The Protestant community is further divided into two denominations, Church of Ireland and Presbyterian, each with its own church. There are therefore three churches in Rathbeg. There are two Primary schools and the nearest second level schools are in a town twelve kilometres away. A community hospital in a nearby village provides residential, respite and day care for frail and incapacitated older people in the environs of Rathbeg. There are private nursing homes in the principal towns in the county.

Social change in Rathmore and Rathbeg

While Rathmore and Rathbeg appear to represent opposite ends of the rural-urban continuum we found features of traditional rural living in Rathmore and features associated with urban living in Rathbeg. For example in Rathmore we found religious practices and forms of neighbouring reminiscent of communal traditions in rural Ireland. We also found close long-standing relationships among neighbours who had for example been the first occupants of a new housing estate. In contrast, Rathbeg has experienced many changes that some older residents perceived as 'losses', such as a decline in the traditional cultural practice of neighbours visiting each other. It seemed clear that the significant relationships, interdependencies and commitments that we set out to investigate in Rathmore and Rathbeg would not correspond with simple notions of traditional and urban ways of life. For this reason, it was important to take account of the articulation of social life in both localities with social linkages beyond their boundaries.

Outline of book

Chapter One sets out the aims and rationale of the study. The types of relationships and activities that are the focus of the investigation into the communal lives of older people are explained: these include leisure and social activities, club involvement, religious practices, volunteering, informal helping, involvement in community groups and informal interactions. The chapter gives a flavour of the themes to be explored and the methods used in the study. It describes the two localities chosen to investigate processes related to ageing and community, Rathmore in Dublin and Rathbeg in County Donegal.

Chapter Two explores changing understandings of old age, and ageing in industrialised societies generally, and in Ireland in particular, over recent decades. The chapter examines evidence of generational relations, the social and economic role and cultural attitudes towards older people in Ireland in the past and in recent times. Contemporary social theories on ageing are outlined and compared including political economy, life history, developmental and humanistic theories. A humanistic approach to understanding ageing is the approach adopted in the book and arguments are advanced that ageing must be understood in relation to community. Both traditional and contemporary accounts of community are examined and a critique is made of some theories that focus on personal communities and that eschew the significance of communities of place.

Chapter Three examines the view of the older person contained in social policies in Ireland, the practical effects of policies for the well-being of different groups of older people and the extent to which social policies enhance possibilities for continuing personal development and cultural and social integration. The evolution of social policy is traced through key reports from the 1960s to date. Principles that have guided policy including the idea of a continuum of care, 'ageing in place' and healthy ageing are discussed and critiqued. The difficulties that older people experience in day-to-day living are discussed on the basis of accounts given by older people in the two localities. The chapter argues that the services and supports necessary to give effect to principles of 'ageing in place' have been slow to develop. Nonetheless, examples are given of good locally-based day services that provide a range of supports and opportunities for sociability and creativity.

Chapter Four examines social participation through leisure activities, social clubs and informal sociability. The chapter maps out the most significant social groupings in the lives of older people in Rathmore and in Rathbeg. They include sports groups, social clubs, Active Retirement groups (ARAs), special interest groups, voluntary and community groups and church groups. The leisure activities and interests of older people in Rathmore and Rathbeg are described. Patterns of participation in social and leisure activities are discussed highlighting differences in terms of gender, age, social class and location. The social and personal significance of leisure activities is examined, drawing on theories and concepts related to lifespan development and the Third Age. The overall argument in the chapter is that through communal leisure activities people create social worlds that make life meaningful and worthwhile.

The theme of Chapter Five is the meaning of place in Rathmore and Rathbeg. It explores the respondents' attachment to the places where they live. Settings where meaningful communal engagement takes place are described and the different ways in which connectedness is created are explored. Data is presented on the likes and dislikes of older residents about their community and on their ideas about older people's involvement in their community. Social class and gender differences in relation to the experiences of neighbouring and community are also discussed.

Chapter Six examines volunteering, informal helping and religious practices among older people in Rathmore and Rathbeg. The theme of this chapter is positive neighbouring as an important social resource. The strong ethos of practical help among neighbours is described. Religious practices and the meaning of religion are also explored. Patterns of churchgoing are presented, as well as evidence on the personal and social significance of religious involvement. Activities such as volunteering and informal helping involve creating a more habitable social world, and voluntary and church-based groups play an important role in facilitating this.

Chapter Seven examines the diverse patterns of relationships between groups of older people and their families, kin, friends and neighbours. It also explains the complex ways in which kinship and community can be experienced in the older years. Data is presented to delineate patterns of connectedness, sociability and solidarity in the lives of older people in Rathmore and Rathbeg. From this data it is possible to compare urban and rural experiences, and to examine the experiences of older people who have extensive kin and community relationships as well as those who experience loneliness or social isolation.

In Chapter Eight a model of connectedness is presented based on both qualitative and quantitative data. The model which illustrates how older people live in their communities is supported by theoretical and empirical arguments drawn from the study. Five indicators of connectedness are proposed and a typology of ten relational types is developed. The typology takes account of levels and degree of contact with family, kin, neighbours and friends as well as participation with others in their community through leisure, voluntary work and informal helping. The chapter also discusses factors that are strongly correlated with high and low levels of connectedness.

Chapter Nine explores friendship in socio-spatial settings in general and in the study areas in particular. The types and contexts of friendships in Rathmore and Rathbeg are described. Friendships observed include: one-to-one friendships that are supportive and help people to 'get by'; friendship among neighbours that is often expressed in practical acts of kindness; friendship among members of clubs and social groups that is characterised by sociability and fun; and generalised friendliness that involves acknowledging, greeting and being interested in the lives of others in one's locality. We question ideas about friendship that overemphasise psychological closeness, choice and individuality. We argue that traditional accounts of friendship that stress sociability and moral obligations are closer to the types of friendship we observed among older people in Rathmore and Rathbeg.

Chapter Ten draws together the evidence on the social integration, community involvement and satisfaction with life among older people in Rathmore and Rathbeg. The study depicts a relational landscape made up of dense interactions and highly meaningful everyday activities. We contend that these realms of social activities can seem in danger of being ignored in Ireland at the start of the 21st century. Policy implications discussed here relate to the positive contribution of older people themselves to social and public life and the importance of recognising the natural solidaristic relationships within localities. The chapter revisits the concept of community and proposes that community implies the possibility of rich multi-dimensional relationships, many of which are geographically based. Lastly, we argue that older people are particularly well placed to build up this type of community and to provide its resources to others.

Chapter Two
Context of Ageing and Old Age in Ireland

Introduction

While Ireland was relatively late to develop industrially compared with other European countries, Ireland has experienced sudden and sustained economic growth and prosperity since the mid-1990s. We have already commented on the mix of traditional and modern values that are found in this multi-layered and rapidly changing society. While diversity and pluralism are now more evident, religious practice is strong in Irish society in comparison with the UK. Another cultural difference between Ireland and the UK relates to property ownership and the meaning of place. Edmondson has described the symbolic meaning of property ownership in rural Ireland in confirming the social identity of family members even when they move away from home:

> In rural societies especially, the 'home place' locates members of the family on a social map which helps to confer on them membership of a community; and the family home forms a focal point in which a couple's adult children, possibly scattered all over the globe, may gather (Edmondson, 2000: 22).

In this chapter we examine ideas about the nature of ageing and the nature of community in a society which is made up of cultural elements associated with the past and cultural influences associated with globalisation and modernisation.

Ageing in Ireland in the past

In order to understand the lived experiences of older people in contemporary Irish society we must examine evidence of ageing in Ireland in the past and consider more recent socio-demographic, socio-cultural and socio-economic trends which have influenced processes and experiences of ageing. The historical and ethnographic evidence that exists on household patterns, generational relations and attitudes to older people in the early decades of the twentieth century show that Ireland was unique in many respects in comparison with much of industrialised Europe.

Patterns of co-residence are of interest to historians and sociologists of the family because of what they tell us about intergenerational sharing of material resources and the socialisation of children. From the point of view of social gerontologists they suggest what social role and influence older people had. Despite some controversy over the extent to which Irish households varied from the nuclear pattern found in much of North Western Europe, Guinnane has found on the basis of data from the 1901 and 1911 censuses that rural Irish households included many kin who would not normally be found in a nuclear family household. Guinnane showed that while the co-residence of two married couples of different generations was rare the dynamic characteristics of a stem family system, such as transmission of the property to a son who marries and remains in the household, was a central feature of Irish households in the early decades of the twentieth century (Guinnane, 1996: 102). This can be contrasted with the nuclear family structure which became dominant and in which most older people now live with their spouse or alone if they are widowed or with unmarried children.[1]

---

1 A large randomised survey of older people in the Western Health Board and Eastern Region Health Authority areas found that the most common living arrangement was for older people to live with one other person, usually a spouse (30 per cent) while 28 per cent lived alone (Garavan *et al.*, Health and Social Services for Older People report, 2001).

Arensberg and Kimball noted that rural Ireland in the 1930s had become 'in some ways an old person's country' (Byrne *et al.*, 2001: 153). They suggested that older people, especially males, were accorded unusual respect and social authority:

> In the Irish case we shall see that they live long because they have much to live for. In their own sphere of life they are honoured. They have power ... Place of honour and privilege within the household is theirs. In their families they are objects of respect and a mild sort of veneration on the part of the younger members (Arensberg and Kimball, 1968: 162, 163).

In some respects their power and authority stemmed from their control over the family farm. The social position of adulthood was associated with farm ownership and young men were referred to as 'boys' regardless of chronological age so long as their fathers still ran the farm. Even in social conversation or representations to outside bodies the older men were expected to be the leaders (Arensberg and Kimball, 1968: 171). A huge turning point occurred when the land was transferred usually to the eldest son.

While the status of older farm owners was high the status of single older people was less certain. The low marriage rates that prevailed in Ireland in the 1930s, 1940s and 1950s meant that a large proportion of those entering adulthood in that period did not form procreative families of their own. In 1991, for example, 23 per cent of the population aged over 65 in Ireland was unmarried which was the highest such proportion in any western country (Fahey and Fitzgerald, 1997: 81). In the past the single elderly accounted for a disproportionately large share of the elderly in long-term residential care (National Council for the Elderly, 1985: 23).

Family formation patterns and kin networks

The main features of the family formation patterns in Ireland from the 1930s until the 1960s (the period when the older people in this study were in the family formation stage of the life cycle) were an exceptionally high incidence of non-marriage, high rates of emigration until the 1960s, high rates of migration to urban areas within Ireland, high marital fertility until the 1960s and improvements in survivorship throughout the twentieth century. The evidence from Irish research has confirmed the existence of high levels of contact with kin across classes (Whelan and Whelan, 1988; Fahey and Murray, 1994; Layte *et al.* 1999). Fahey and Murray concluded that the majority of older people at the period when their study was carried out were likely to have large numbers of lateral and descendent kin while a sizeable minority who never married would have significantly smaller networks of relatives (Fahey and Murray, 1994: 100–103). Fahey and Murray found that while both married and widowed older people had extensive kin networks both locally, in Ireland and abroad, the unmarried elderly, both those living with siblings and those living alone, had much more limited kin networks including fewer siblings (Fahey and Murray, 1994: 107–110). In an examination of welfare implications of demographic trends Fahey and Fitzgerald suggested, however, that this maldistribution of the social resources which families represent in the elderly population will reduce in the short to medium term (Fahey and Fitzgerald, 1997: 94). Furthermore, there is projected to be a sharp decrease in the proportion of the older population who are single and an increase in the number of older people living alone.

Present and future demographic trends

In terms of the age distribution Ireland is a young country in comparison to its European neighbours (O'Shea, 2006). At the time of this study there were 435,000 persons aged 65 and over in Ireland, representing 11.2 per cent of the total population[2] (Central Statistics Office, 2002). Population projections suggest that older people will form a larger proportion of an expanding population over the next two decades. It is projected that, by 2031, the proportion of older people in Ireland will have almost doubled to between 837,000 and 858,000 people, representing between 18 and 21 per cent of the population

2 In 2006, Ireland had the lowest proportion of its population aged 65 and over among EU countries at 11 per cent, this compared to an EU 27 average of 16.8 per cent (Ageing in Ireland, 2007: 10).

(DOHC, 2001). The marital status and household composition of the older population are also projected to change substantially over the coming decades. The marriage rate rose sharply in Ireland in the 1960s and 1970s and it is projected that the largest increases in the older population in the coming decades, both in absolute and relative terms, will occur among married men and women. There will be a sharp decline in the proportion of older people who are single.

Based on 2004 population projections there will be an increase in the population aged 65 years and older of about 60 per cent between 2002 and 2021. There will be a general shift towards married and separated older people and a shift away from single and widowed. The proportion of married people in the older population will rise from 49 per cent to 60 per cent between 2002 and 2021 while the proportion who are single will decrease from 18 per cent to 11 per cent (Connell and Pringle, 2004). The proportion who are widowed is expected to fall from 33 per cent to 28 percent.

The expectation of a large increase in older people who are divorced already documented in the UK (Arber *et al.*, 2003) is likely to be an emerging trend in Ireland. While marital separation is currently very uncommon among older age groups, Connell and Pringle suggest that the relatively high separation rates of those in their late forties and early fifties evident in 2002 will have fed through to much higher rates of separation in older age groups than is currently the case, rising to 8 per cent of the older population (Connell and Pringle, 2004). There is also a growing propensity for older people to live alone. Up to 2006, the growing propensity of the single and widowed to live alone was cancelled out by the growing proportion of older people who are married. However, the overall number of older people living alone is predicted to almost double between 2002 and 2021, with the proportion living alone increasing from 25 per cent to 30 per cent (Connell and Pringle, 2004).

Social theories of ageing

A wide variety of theories and concepts in social gerontology have been used in recent decades to interpret ageing as a socially constructed phenomenon. While no composite theory of ageing may be possible given the complexities involved in processes of ageing, different theories provide particular insight into aspects of ageing. In the present study we need theories that apply to the different types of individuals being investigated – women and men, those without property as well as those who are well off, those who come from urban backgrounds as well as rural, in short, theories that take account of the different life histories and life chances of a diverse group of older adults. Social theories of ageing vary in the degree to which they are prescriptive or neutral in relation to moral and political arguments. For example life history approaches recognise that particular groups of birth cohorts would have experienced a particular configuration of economic, political and social circumstances that would have circumscribed their life chances, yet, they often remain neutral as regards the political or moral implications of such inequalities. On the other hand political economy theories offer analyses that carry clear public policy prescriptions. Like other social theories, theories of ageing are subject to changes in the light of the broad acceptance of ideas about the nature of people and society. These theories are therefore, in part, a matter of academic fashion in the light of dominant political and moral views and as such may be seen as a product of their time.

We will begin by examining approaches that emphasise individual adjustment to ageing. Integration and adaptation were seen by both disengagement theory and activity theory as meeting the needs of ageing individuals and society. However, theories from functionalist perspectives such as that of disengagement came to be criticised for appearing to attribute problems of ageing to individual adjustment as well as for working with an unrepresentative sample and appearing to advise older people to withdraw from society and make room for younger people. Psychological theories emphasise individual development and adaptation to the demands of adult social functioning. Of particular interest from the point of view of the present study is the work of Erik Erikson who focused on the possibilities for development at all stages of the life course – rather than assuming it is allowable to instruct older people how to live their lives rather than, say, 35-year-olds. Political economy perspectives highlight the existence of structured inequalities among older people and generally argue forcefully against them. Life history perspectives suggest the importance of understanding aspects of ageing in the past. In the Irish context this would include the power and authority of male property owners, patterns of marriage, celibacy and intergenerational co-residence, emigration, poverty and large family size, limited educational opportunities and the subordination of individual's life plans to suit the needs of parents and other family members. Humanistic perspectives focus on the question of what gives meaning in old age. Writers such as Coleman in the UK and Moody in the US have stressed nonmaterial sources of meaning such as religion.

Functionalist perspectives

The American functionalist theory of the post-War period posits a fit between the orderly functioning of society and the 'natural' inclination of individuals to adapt or change their social activities and roles as they get older. During the 1950s the predominance of functionalist perspectives in social gerontology such as disengagement theory and activity theory attributed characteristics to older people that were assumed to represent a state of equilibrium between the ageing individual and society but may in fact have concealed ideological positions that did not reflect the lived experiences of older people. Disengagement theory, originally formulated by Cumming and Henry (1961), stated that, independent of other factors such as poor health or poverty, ageing involves a gradual but inevitable withdrawal by the individual from their previous social activities and commitments:

> Ageing is an inevitable mutual withdrawal or disengagement resulting in decreased interaction between the ageing and others in the social system to which he belongs. The process may be initiated by the individual or by others in the

situation … His withdrawal may be accompanied from the outset by an increased preoccupation with himself; certain institutions in society may make this withdrawal easy for him' (Cumming and Henry, 1961: 14).

The process of disengagement was seen in the loss of roles such as the work role, mainly in the case of men, and the homemaker role in the case of women. The assumption was that both the individual and the wider society benefited from this process of disengagement which involved a release from social pressures and stress for the individual and an orderly transfer of responsibility and power from one generation to the other. While ageing is indeed associated with processes such as retirement and disability, and with transitions involving loss of significant relationships because of death and less intensive parenting responsibilities, disengagement theory has been widely criticised. The grounds for such criticism include that it is overly simplistic, it makes implicit value judgements that marginalise older people, it adopts a biomedical model of ageing and there are serious methodological flaws in the empirical work. It has been pointed out, for example, that people are adaptive in substituting different forms of interaction for former ones and that people react in a variety of ways to retirement and loss of former functions, including in some cases seeking to continue a particular pattern of life. Furthermore, conceptualisation of old age as a time of passivity and withdrawal has been challenged as not being in the interests of either older people themselves or of society. Finally, it has been argued that the methods used by Cumming and Henry to obtain their sample, using 'snowballing techniques' resulted in a homogeneous and unrepresentative sample of middle class white Americans.

Activity theory, developed by Havighurst (1963), portrayed an alternative view of ageing, maintaining that normal and successful ageing involves preserving the attitudes and activities of middle age for as long as possible. In many ways this theory foreshadowed ideas about the Third Age of activity and leisure that were popularised in the 1970s and that have challenged assumptions that withdrawal of older people from employment or active work is either personally or socially beneficial. Activity theory may also be criticised, however, as being over simplistic and prescriptive by suggesting that everybody can aspire to continuing an active life style associated with middle age. In this sense activity theorists may be denying old age altogether.

The Third Age

A new stage in conceptualising significant divisions in the life course, The Third Age, has been suggested as marking the period when an individual has retired from formal work but is physically and mentally active and able to engage in active leisure (Laslett, 1996). The postulation of this stage reflects a trend towards early retirement, an earlier release from parental responsibilities due to declining family size and the increased affluence of many middle-aged, middle-income groups. Laslett suggests that the Third Age made an appearance in Britain during the 1950s and began to establish itself as a settled feature of the social structure during the 1980s.[3] (Laslett, 1996: 101). Laslett used the term the Third Age in his first edition of *A Fresh Map of Life* to introduce a new life-course stage, marking it out as a crucial stretch of life and differentiating it in particular from the Fourth Age which was associated with illness and decline (Laslett, 1989).

Underlying Laslett's views appears to be a vision of the purposes of life as being about personal development involving intellectual learning over the whole of the life course and education as a value in itself, and emphasising duties as well as privileges. To what extent does Laslett's view reflect the reality of older people's lives? One criticism of Laslett's theory is that it is rather elitist in its assumptions that people in the Third Age are interested in intellectual pursuits and education. Indeed it can be argued, and Laslett himself acknowledges this, that many of the possibilities that Laslett believes are open in the Third Age, such as getting involved in the University of the Third Age

3 Laslett suggests that the term the Third Age entered the vocabulary in the 1980s and was originally used to describe the new educational societies, Les Universities du Troisieme Age established in France and Spain in the 1970s (Laslett, 1996: 3). The Third Age has been defined for statistical purposes as expected between 50 and 74 (See for example the Carnegie Enquiry into the Third Age). The Third Age appears to fuse the later part of non-old adulthood and the active younger part of old age (Laslett, 1996: 1x,x).

or helping to run heritage institutions, require a high degree of education to appreciate and act upon them (Laslett, 1996: 265).

Allied to claims that his ideas are elitist is the argument that most people do not see their lives as planned projects and are therefore unlikely to undertake the planning that Laslett suggests is necessary in the Second Age to ensure that the Third Age is a period of productive and creative pursuit of interests. A further difficulty with Laslett is his conceptualising of the Fourth Age. Laslett is clear that his purpose in separating out the Third Age from the Fourth Age is to disassociate becoming older from negative connotations of inactivity, exclusion and diminished status (Laslett, 1996: 5). However, his wish to promote a positive image of the possibilities in the Third Age may have led him not to look for life affirming possibilities in late old age when physical decline is inevitable.

If we consider many of the ways in which public arrangements for older people are applied it often appears as if disengagement theory had united with a biomedical emphasis on the problems of old age leading to generally dispiriting results. A focus on problems and deficits of older people and assumptions that older people should not expect more from life than care and maintenance seems to have pervaded much public policy provision for older people (Gallagher, 2005). For example in extended care facilities for older people, routines are often designed for efficiency from the organisation's viewpoint and there has been insufficient attention to individuality, self-autonomy or normal daily living. By contrast, the beneficial social consequences for older people of adopting ideas about an active old age, that appears to fit better with activity theory, can be observed in initiatives to introduce more purposive leisure activities into residential and day settings where older people interact. At the same time our understanding of the lived experiences of older people must not ignore the reality that some degree of withdrawal from activities is inevitable, especially where people live into advanced old age.

Yet as individuals age they retain most of their social and political responsibilities as distinct from work responsibilities. Older people of any age contribute to the quality of life of family and kin through being involved in a long-term chain of support and in many instances practical help (see for example Finch, 1989, and Arber and Attias-

Donfut, 2000). At the same time due to frailty and incapacity some are not able to give practical help and indeed depend on others for many of their own day-to-day needs.

Psychological perspectives: life span and life stages

Life span theories conceptualise the ageing process in terms of more or less distinct phases from childhood through adulthood to old age, premised on the idea of fixed developmental tasks and social roles appropriate for each phase. Erikson's theory of personality development provides a framework for the whole lifespan in terms of psychosocial stages involving developmental tasks which are appropriate for each stage but which may be resolved or unresolved to varying degrees as individuals move through the life cycle (Erikson, 1980). According to Erikson each stage involves trying to achieve a balance between opposite tendencies, for example the task of the last stage of life is to attain 'ego integrity', 'an assured sense of meaning and order in one's life and in the universe, as against despair and disgust' (Erikson, 1980). Erikson and his colleagues applied the ideas of 'vital involvement' and 'involved disinvolvement' to interpret and to understand the tasks and challenges of old age:

> The eight stages of psychosocial development, which produce the strengths necessary for mutual involvement in an ever-increasing social radius, from infancy through adulthood, and into old age. And this is our basic interest in this study: the remaining or new potentials of the last interactions and the vital (if paradoxical) involvement in the necessary disinvolvements of old age (Erikson, *et al.*, 1986: 33).

Where the authors perceived adulthood to mean a set of vital involvements in life's generative activities including positions of responsibility in the family and community, 'involved disinvolvement' was seen to characterise old age. They represented this idea by the role of grandparenthood that involves an older person in a different type of caring/nurturing role to that of parents and 'which must remain loyal to a defined and planned role for old age within an order of wisdom'

(Erikson *et al.*, 1986: 50–51). Indeed Erikson and his colleagues define wisdom as true involved disinvolvement:

> Wisdom is detached concern with life itself, in the face of death itself. It maintains and learns to convey the integrity of experience, in spite of the decline of bodily and mental functions (Erikson *et al.*, 1986: 37–38).

Another essential aspect of involvement is 'heightened awareness' – the final maturation of an existential identity as well as confronting one's death (Erikson *et al.*, 1986: 51). This is explained by the two essential qualities that the authors suggest must characterise vital involvement in old age, actuality and mutuality. These qualities are understood to mean 'a sense of reality that is an active and, in fact, an interactive involvement within a communal "actuality", and a shared sense of "we" within a communal mutuality' (Erikson *et al.*, 1986: 53).

The very concepts used by Erikson *et al.*, while complex and challenging, appear to resonate with the lives of older people in Rathmore and Rathbeg. Vital involvement may mean engagement with other people in one's neighbourhood, participating in mutual social and leisure activities and seeking continuing opportunities to experience new things. Thus involvement in clubs and groups, voluntary work, hobbies and adult education classes may all contribute to vital involvement. The idea of 'involved disinvolvement' refers to norms in most western societies that one is no longer at the front line of the work organisation or active parenthood but plays a somewhat less involved role, for example semi-retired, unpaid work or active grandparenting. Erikson and his colleagues suggested some potential roles that older people could have in society, for example, acting as conservationists, being arbiters in disputes and involvement in the Arts and lifelong learning.

Erikson's understanding of the tasks of old age helps us to see older people in the context of their whole life history with problems both successfully and unsuccessfully resolved from earlier periods in life. Bond *et al.* point out important implications of this including:

1. that people diverge more and become more unique as they accumulate life experiences,

2. that each stage of life has its own essential features and possibilities for development, and
3. that there are reciprocal influences between the individual and his environment including the influence of the young on those who are older as well as the more obvious influence of parents and society on the young (Bond *et al.*,1993: 30).

However, what stands out in particular from Erikson's work is the extent to which it addresses questions of meaning in later life and the importance of the spiritual dimension in understanding old age. A spiritual dimension is understood as the idea that life has a transcendent aspect – that there is a reality beyond our own inclinations and socialised being. It refers to belief in a higher power and the possibility for humans to develop a higher awareness and a higher capacity to love God and one's fellow humans. Awareness of a meaning that transcends individual life is a universal human quality. Erikson *et al.* explain that heightened awareness characterises vital involvement in old age and involves both acceptance of one's finite nature and a sense of common humanity:

> One must face oneself as one who shares an all-human existential identity, as creatively given form in the world religions. This final 'arrangement' must convince us that we are meant as 'grandparents' to share the responsibility of the generations for each other. When we finally retire from familial and generational involvement, we must, where and when possible, bond with other old-age groups in different parts of the world, learning to talk and listen with a growing sense of all-human mutuality (Erikson *et al.*, 1986: 337).

Thus Erikson and his colleagues appear to be more inclusive in their approach than say Laslett in seeing the possibilities for 'vital involvement' in both active and inactive old age through engagement in reciprocal intergenerational and communal relations. Erikson used the expression 'grand generativity' to express the idea of older people showing sincere and effective concern for others. (Erikson *et al.*, 1986: 178). Therefore, regardless of the degree of incapacity an older person may be experiencing, through concern for the welfare of others and of future generations older people can find a sense of meaning in life.

Political economy theories

Political economy perspectives reflected in the work of Estes (1979, Estes *et al.*, 2003), Walker (1981, 1993), Townsend (1981, 1986) and Phillipson (1982, 1998) suggest that how older people experience life particularly in relation to incomes and security is determined largely by their economic and social status preceding retirement and to political decision making in welfare capitalist societies. These decisions determine, for example, the circumstances of exit from the workplace, the rates of state benefits and pension arrangements. Political economy perspectives have suggested that:

> Dependent identities in later life are not an inevitable and biologically or developmentally determined 'fact of life', but are a social artefact produced by a number of factors such as: the forced expulsion of older people from the workforce; the growth and self-promotion of professional interests; the manipulation of vulnerable older people as if they were commodities in the welfare market; and a marginalization of later life concerns (Estes *et al.*, 2003: 29).

The concept of 'structured dependency' was first developed by Townsend (1981, 1986) who argued that social and economic relations associated with welfare capitalism restricted access to a wide range of social resources and fostered passivity in old age, a possible explanation for some of the findings of Cumming and Henry. The political economy perspective has long provided a critique of the liberal-pluralist theory in political science and sociology and offers diagnoses and prescriptions clearly based on a critical analysis of economic inequalities in welfare capitalism. It is clear that from diagnoses like these public policy prescriptions are likely to follow. In recent years the concept of 'social exclusion' has been useful in highlighting the social costs that arise when certain groups, including categories of older people, become disengaged from society (Scharf, *et al.*, 2004).

However, a number of criticisms have been made of the concept of 'structured dependency'. It has been suggested that theorists such as Townsend tended to concentrate on the relatively small group of older people living in institutions and did not acknowledge the range of choices made available to older people by virtue of improved pensions, sometimes making them better off than younger people in employment (Johnson, 1987). Townsend has also been criticised for himself attributing passive characteristics to older people and for stereotyping them as 'cultural dopes' simply because they have a reduced functional role in society (Dant, 1988).

On a larger scale, however, the overall effects of changes in capitalism which are evident since the 1970s (what Lash and Urry (1987) have called an end to organised capitalism) are far from clear. Flexible forms of work and de-industrialisation are changing the certainties associated with career and fixed retirement and forcing people to negotiate new choices about how to conduct their lives (Phillipson, 1998: 47). In the UK Phillipson has argued that two institutions that provided frameworks for older people since the period of the Second World War, namely retirement and the welfare state, are under threat and that changing priorities and values associated with a more competitive and individualistic society 'undercut the construction of a viable identity for living in old age' (Phillipson, 1998: 53–54). However, these changes have also been interpreted as providing opportunities for older people to develop satisfying life styles in the middle years which period of time is now defined very fluidly. Featherstone and Hepworth reflect on possible positive benefits: 'this elaboration of mid-life increasingly implies a flexible, individualised, biographical approach which takes into account human diversity (Featherstone and Hepworth, 1989: 154). Yet, even if these changes can be as positive as Featherstone and Hepworth suggest, it is also true that it may only be an affluent group who can avail of new opportunities for personal growth during the older years.

Some writers suggest that the emergence of the Third Age is a transitory stage towards the general transformation of our society 'with older people helping to bring more plural values into the second age and with old and young and all in between combining paid work with unpaid, and with no work in new patterns not yet conceived' (Schuller and Young, 1991: 173). Schuller and Young challenge restrictive conventions about age appropriate roles, regarding age stratification associated with work/retirement norms in capitalist-welfare societies as an artificial device for keeping people in line. The systematic stereotyping of older people or 'ageism' has been challenged through increased political activism on the part of older people and their representative associations and has led to legislative protection in many countries to outlaw discrimination on age grounds (for example in Ireland the Equal Status Act, 2000).

In relation to older people in this study, many of the changes associated with ideas of the Third Age and choice in relation to manner and staging of retirement may have come too late for them, or they may not take identical form among older people in Ireland or they may apply only to a certain socio-economic grouping. We will examine Irish social policy responses towards older people in Chapter Three. Notwithstanding these qualifications, we would argue that the general points of political economy perspectives stand; quality of life including involvement in satisfying leisure and development of interests takes place within contexts strongly affected by socio-economic structures.

Life history approaches

A serious limitation of many theories of ageing is that they are culturally and historically specific and they often expressed moral and political approaches to which critics took exception. Life history approaches try to be less prescriptive by taking the wider social context into account and recognising that birth cohorts would have had shared experiences of a particular configuration of economic, political and social circumstances unique to their time. A life history approach has been used effectively to give insight into processes of social reproduction. Bertaux used the method of collecting life story interviews to understand the social relations of production organising the process of bread production in the family bakery in France in the 1970s. Not only did this method provide insight in to the process by which the artisanal bakery reproduces itself over successive generations but also contributed to theoretical and methodological development that was critical of functionalism and structuralism (Bertaux, 1982).

In the case of the participants in the present study the vast majority would have been born between 1917 and 1937. Habits and expectations would have arisen from a period when there was great political upheaval, the Irish State was formed with a deeply conservative ideology and there was no money to spend on social services (Burke, 1999: 2526). Ethnographic accounts of rural Ireland in the years between the Two World Wars paint a picture of a society with many 'traditional' features:

> Small farm society in the inter-war years was remarkable for its localistic frame of reference, acceptance of low living standards and adherence to 'traditional' patterns of economic and social conduct (Byrne *et al.*, 2001: LXV11).

The older people in this study would have reached the legal age of adulthood – then 21 years and have begun to form their own families during the decades of the 40s and 50s. Conroy has described the decade of the 1950s as a period of economic stagnation, high unemployment and mass emigration (Conroy, 1999: 34). This period provided conditions that made it extremely difficult for significant numbers of people to obtain economic security in their own country or to form procreative families of their own. While Ireland remained neutral during the Second World War, many Irish people enlisted with the British army or worked in Britain where there were job opportunities during the war. While it is not possible to generalise about the precise expectations of two entire 10-year cohorts born in those times, we should be alive to the possibility that traces have been left in the ways of life led by individuals now living.

Life course perspective

A life course perspective provides both a developmental and historical framework for the study of intergenerational relations. It involves three dimensions that revolve around timing: individual timing of life transitions in relation to external historical events; the synchronisation of individual life transitions with collective familial ones; and the impact of earlier life events, as shaped by historical circumstances on subsequent ones (Hareven, 1996: 3). It was not uncommon in the past and is still the case in many societies that individual life trajectories depend on collective family needs. For example Hareven and Adams found among several ethnic groups in the United States at the turn of the twentieth century that ageing parents often discouraged at least one caretaking daughter from leaving the parental home and marrying in order to provide continuing support in old age (Hareven and Adams, 1996).

We too found evidence that, while emigration was a common experience for many of our respondents in young adulthood, the decision to emigrate had to be synchronised with the needs of families. We came across cases where unmarried daughters were dissuaded from emigrating in order to care for their ageing parents and subsequently cared for their unmarried brothers on the family farm after their parents had died. We noted that large families and low incomes were commonplace and our respondents told us that the education of children was accorded low priority in many families. One respondent remarked that she was taken out of school every time her mother had a baby which had a detrimental effect on her ability to read and write, something she felt ashamed of in later life. Many of our respondents started work as young as fourteen years, sometimes due to the early death of a parent. One respondent in Rathbeg had worked from the age of seven herding animals. He went to school at the age of eleven because compulsory education was introduced at that time and he left again at fourteen.

It has been suggested that life course perspectives permit a more dynamic approach to relations between the individual, the family, work and others, and highlight the capacities of members of differing generations to sustain reciprocal relations over time (Featherstone and Hepworth, 1989: 154). Arber and Evandrou advocate a life course perspective as contributing to a more dynamic view of ageing which highlights the agency of older people themselves but also shows how their diverse biographies are affected by their gender, class and ethnicity (Arber and Evandrou, 1993). Thus what the life course perspective contributes is an understanding of the different social worlds that older people inhabit simultaneously, as a result of having lived through a variety of familial, work and welfare configurations. Patterns of providing support and expectations for receiving support in the older years for example are part of a continuing interaction among parents, children, kin and neighbours shaped by values, experiences and institutional norms over the entire life course.

Our present cohort of older adults have benefitted from the substantial improvements in state pensions and social security arrangements, particularly during the decades of the 70s and 80s. Whereas in the early 1970s the risk of income poverty was high for older people relative to other segments of the population, by the late 1980s their position had improved both absolutely and relatively (Layte *et al.*, 1999: 16). Indeed many of our respondents expressed the view that there was generous provision of health and welfare benefits for older people in Ireland today.

However we must be cognisant of the very different experiences of subgroups within age cohorts. Those without property or education in the early decades of the twentieth century in Ireland would have experienced forced emigration and harsh working conditions that would have affected their socio-economic position in later life in terms of home ownership and pension provision. We came across individuals who appeared to have experienced cumulative disadvantages, including involuntary emigration, poor working conditions and disability, sometimes as a result of occupational injury. In designing our study it was important, therefore, to take account of the factors that influence the possibility of having a comfortable old age, (for example as a retired home owner with a relatively good pension), more easily than experiencing material disadvantage (for example retired early due to disability or forced redundancy, living in public accommodation and depending entirely on the state old age pension).

Humanistic approaches

Humanistic perspectives in social gerontology focus on meaning in old age and offer insights into the values and moral questions around what makes life satisfying and worthwhile. How individuals experience their lives as meaningful in their older years has been addressed by philosophers and writers down through the ages. Cicero in his classic essay, 'De Senectute' defined what he saw as a good life in old age. The idea of a good life for older people has emerged in recent public policy discourses in the form of ideas about 'successful ageing' or 'ageing well'.

Phillipson has suggested that a crisis of meaning is facing older people in the post modern world because the frameworks within which people have been traditionally anchored have become problematic (Phillipson, 1998: 52). While Phillipson's analysis of the search for meaning in old age relates mainly to changes in the material conditions of life in old age that may affect people's sense of inclusion and security, other writers have highlighted that old age is a time when we are likely to come face-to-face with questions of ultimate meaning. Coleman has investigated the importance of religion and spirituality for both health and well-being and for providing a sense of coherence and meaning in old age (Coleman, 2006). Moody suggests that there are two questions to be asked about meaning in old age: firstly, does old age have meaning for society? and secondly, how do individuals experience their lives as meaningful in the last stages of life? (Moody, 1998: 438).

Cicero's view of old age

For Cicero, continuous learning, culture and living a virtuous life were the means to secure a happy old age:

> The arms best adapted to old age are culture and the active exercise of the virtues – the harvest they produce is wonderful, not only because they never fail us even in our last days (though that in itself is supremely important), but also because the consciousness of a well-spent life and the recollection of many virtuous actions are exceedingly delightful (Cicero, para 3).

Cicero discussed four reasons for old age being thought of as unhappy; it withdraws people from active employment, it enfeebles the body, it deprives people of nearly all physical pleasure and it is the next step to death (Cicero, para 5). However, despite inevitable losses of physical strength and sensual pleasure Cicero sees the cultivation of an interior life as not only offering compensation for these losses but enhancing satisfaction with life itself. For Cicero being active and pursuing interests were the key to a satisfying old age. For example he saw farming as a satisfying occupation in old age and gave eloquent accounts of the joy of observing nature and cultivating the land (Cicero, para 15).

In reply to criticism that his prescriptions applied only to men of high rank and wealth Cicero argued that all older men could keep their minds active:

> Old men retain their intellects well enough, if only they keep their minds active and fully employed. Nor is that the case with men only of high position and great office: it applies equally to private life and peaceful pursuits (Cicero, para 7).

Cicero saw a particular role for older people in providing advice and influence:

> The great affairs of life are not performed by physical strength, or activity, or nimbleness of body, but by deliberation, character, expression of opinion. Of these old age is not only not deprived, but, as a rule, has them in a greater degree (Cicero, para 6).

Moody suggests that Cicero was one of the first and most eloquent proponents of the ideal of 'successful ageing'. Moody suggests that Cicero offered a realistic account of both the gains and losses of ageing and viewed old age not exclusively as a time of decline or loss but also as an opportunity for cultivating compensatory wisdom (Moody, 1998: 81). Cicero's ideas may be somewhat elitist which would not be surprising given his own rank, learning and achievements. Nonetheless, his ideas about the possibilities of older people exercising a positive influence over others and of the old teaching the young has societal implications in terms of how both knowledge and old age are regarded. Cicero's depiction of old age is very positive. He depicts in his writings and in his own life the possibilities of a good life in old age based on the cultivation of life-long interests and values.

Moody has argued that there is a danger of loss of meaning in old age when the self is detached from any larger purpose: 'this deeper dimension of life, a dimension of meaning, is to be found … in those cultural forms – of ritual and continuity between generations – that give intelligibility to the last stages of life' (Moody, 1988: 238–239). Moody examined two domains of meaning: those of religion and leisure activities. Religion plays a vital role in the lives of older people in many cultures. However, as Coleman as pointed out there has been a surprising neglect of religion and spirituality by gerontologists, particularly outside the United States (Coleman, 2005). Religion has traditionally been thought to be especially relevant to adaptation to old age and to coping with losses associated with ageing such as bereavement (Coleman *et al.*, 2002). However, it has been assumed by many writers that modernisation processes inevitably involve an inexorable move towards secularisation in the western world and that there has been a diminution of the salience of religion in people's lives. At the same time spirituality has emerged as a separate subject of investigation from religion and is increasingly contrasted positively with religion, especially in the UK literature. However, Coleman found that for older people in particular talk of the spiritual is intimately associated with religion: 'Even getting them to speak about spiritual beliefs outside of the context of religious beliefs is not easy, although not impossible' (Coleman, 2006).

While there are many different definitions of spirituality and religion, one distinction commonly made is that spirituality involves seeking for meaning outside traditional forms of authority while religion involves acceptance of traditional forms of authority (Coleman, 2006). A longitudinal study of older people in California examined the consequences of being religious and being spiritual. Religion which was defined by belief in God, prayer and attendance at communal worship related to positive relations with others, involvement in social and community life tasks and ratings of generativity. Spirituality which was defined by non-tradition-centred beliefs, sense of connectedness with sacred other and involvement in practices such as meditation related to ratings of personal growth, involvement in knowledge building life tasks, and wisdom (Wink and Dillon, 2007). Wink and Dillon's study suggests that religion is closely related to connectedness with other people and sense of community while spirituality is closely related to personal development and inner awareness.

Coleman has reviewed the research findings on the connections between religious beliefs/practices and health and well-being in later life. The benefits are said to include prevention of physical and some mental illnesses, benefits to longevity, better recovery from illnesses and better tolerance of pain. Possible explanations that have been proposed for these associations include:

1. Religious teaching encourages better health behaviour,
2. Belonging to a religious community provides social support in times of difficulty, and
3. Religion provides a sense of coherence and meaning. (Coleman, P., 2006).

Religious belief and practice can thus be seen as a resource for the individual and the group and may be an important element of connectedness.

Whether expressed in more conventional religious terms or in non-religious terms the search for awareness about themselves and the world often becomes more heightened as people grow older. This search can be described as a search for authenticity: finding what is of most value in one's life, seeking inner contentment, solidarity with others, peace of mind, or resolution of past difficulties. It can also be described as a search for the transcendent: relationship with God, awareness of God's presence, an inner prayerful, contemplative experience. Whatever way older people might express their religious or spiritual needs and aspirations, and leaving room for the often inexpressible part of the total human experience, it can be argued that both religion and spirituality, as understood from Christian and humanistic perspectives, are embedded in people's everyday lives, in their relationships and communities. Personal and social values, often grounded in religious beliefs, are crucial to how older people live their lives. Moral and religious frameworks will permeate the types of interactions and activities that are of value in the lives of older people.

Therefore, the everyday social interactions and relationships that older people are involved in reveal moral frameworks and cultural practices that make life meaningful, tolerable and satisfying.

We have examined a number of theories and concepts that appear relevant in understanding the lives of older people. In the process of engagement with older people these ideas guided our interpretation. Life history perspectives suggest the importance of understanding aspects of ageing in the past in Ireland, including the power and authority of male property owners, patterns of marriage, celibacy and intergenerational co-residence, emigration, poverty and large family size, limited educational opportunities and the subordination of individual's life plans to suit the needs of parents and other family members. When we combine Hareven's understanding of how history shapes biography for cohorts of individuals with political economy perspectives on the inequalities in life chances for individuals within the same age cohort, we can better understand how it is more likely to become one kind of person rather than another in old age. At the same time Erikson's work on the common tasks associated with old age to achieve a balance between integrity and despair through 'involved disinvolvement', mutuality of experience and 'grand generativity' suggest the importance of recognising that all older adults must continue to engage in reciprocal intergenerational and communal relationships.

The concept of the Third Age does elucidate the lifestyles and values of the 'modern retiree' associated with more flexible forms of work, increased longevity and negotiation of new more individualised choices of life style. However, we question the extent to which Laslett's theory of the Third Age may represent an elitist view of life, and may wrongly assume that most people adopt a planned approach to their lives. We also question Laslett's rather negative and limited understanding of the Fourth Age as representing 'an era of final dependence, decrepitude and death' (Laslett, 1989: 4) and seeming to overlook the possibilities for meaningful engagement and life enhancing involvement in late old age and declining health.

How individuals experience their lives as meaningful in their older years and the idea of a good life have re-emerged as important themes in the social gerontological literature. While writers such as Phillipson have expressed concern about the undermining of institutions such as the welfare state that have provided a moral framework of certitude, security and social recognition for retired older people, more recent humanistic approaches have stressed nonmaterial sources of meaning such as religion and purposive leisure.

The question of what gives meaning to lives in old age is an underlying theme throughout the book. The particular focus that we develop in later chapters is exploring the nature of reciprocities and friendship in the lives of older people and we argue that, as Aristotle did, human flourishing takes place within communities. In our examination of the lives of older adults in Rathmore and Rathbeg we examine both the actuality of individual lives in different circumstances and the possibilities of their full cultural and social integration into society.

The nature of community

The extent to which people are embedded in local social relationships or more widely dispersed ties has been conceptualised through concepts related to kinship, community and social network. Community can be thought of as a sociological variable and as a morally valued way of life (Calhoun, 1980: 107). The moral meaning of community was emphasised by early commentators who feared that industrialisation and urbanisation would weaken morals and manners among the working classes:

> Community was moral in that people were not expected to be, in this view, perfect in and of themselves, but rather more perfect as they were better integrated into webs of social commitments, rules and relations (Calhoun, 1980: 106).

In contemporary discourses the moral dimension of community is often seen as residing in communities of interest or communities of identity. Some writers such as Nisbet feared the loss of the 'communal' dimension in modern society and used community to evoke the idea of the person living a more authentic life (Nisbet, 1967: 47–

48). However, in opposition to what they see as a value biased use of the term, many social theorists have eschewed 'community' altogether in favour of concepts such as friendship or social network. We will examine traditional accounts of the nature of community that stress the salience of socio-geographic interdependencies and traditional sources of authority such as kin and church. We will then contrast these with concepts related to social networks and personal communities that purport to better represent the nature of contemporary social relations and plural identities.

Traditional understandings of community

An early and influential concept of community was introduced into the literature by Tonnies in the late 19th century through his writings on the differentiation between Gemeinschaft and Gesellschaft (*Community and Society* by F. Tonnies, first published in 1887 and reprinted in 1967). Gemeinschaft or community he understood as warm, proximate relationships characteristic of village life. In contrast Gesellschaft, usually translated as association, represented the formal utilitarian relationships associated with economic and political life (Tonnies, 1967: 7). Tonnies speaks of three types of Gemeinschaft: Gemeinschaft of blood or kinship, Gemeinschaft of locality or neighbourhood and Gemeinschaft of mind or friendship. According to Tonnies 'neighbourhood' describes not so much a geographical place as the general character of living together in a rural village that involves cooperation in work, in the general running of local affairs and common religious practices. While social relationships may be maintained at a distance through modern methods of communication and frequent travel, symbolic forms are frequently used to reinforce solidarities and friendships such as reunions and other rituals attached to place or family home.

Examples of communal rituals can be observed in two communal events that occur annually in Rathbeg. An annual one-day Regatta has taken place in Rathbeg since 1833. While only a small number of people and very few of the local inhabitants take part in the boat races, the meaning of 'the Regatta' is an annual reunion for old friends and acquaintances who have or have had close associations with the village. Many former inhabitants make a point of trying to return on holidays for the Regatta to meet relatives and renew friendships. However, the event itself has become part of a local festival extending over a longer period that provides a broader range of fun activities for local people and the wider Rathbeg diaspora alike.

Another ritual that takes place in many towns and villages in Ireland is an annual graveyard Mass. Family members gather at the graves of their deceased relatives while an open air Mass is celebrated. In Rathbeg this event is attended by most villagers whether or not they are regular churchgoers. Like the Regatta weekend people come from other localities including a diaspora living away who often plan their holidays to coincide with the graveyard Mass. Family members and other relatives make a special effort to be there indicating that such rituals are important in acknowledging the life of a deceased person and their place within the family and community. These examples seem to resonate with Tonnies' contention that while living in a locality is a general requirement for the type of community he described as 'Gemeinschaft', a sense of community can still be maintained during periods of separation from the locality by 'well-defined habits of reunion and sacred customs' (Tonnies, 1967: 9).

Tonnies regarded friendship as being independent of kinship and neighbourhood and based on similarity of work or common mental and spiritual interests. Friendship is more likely to be found in a town setting since, he argues, the ties of friendship are more easily maintained where frequent meetings are possible and are more a matter of choice than the other forms of community. This, of course, means that Tonnies is not merely drawing a sentimental contrast between old, rural, pre-industrial forms of living together and new, urban, 'modern' life. Tonnies regarded what he saw as the three types of community – kinship, locality and friendship – that are closely interrelated in space and in time as manifestations of the highest form of human life: 'Wherever human beings are related through their wills in an organic manner and affirm each other, we find one or another of the three types of Gemeinschaft' (Tonnies, 1967: 9).

There has been a tendency over the years to associate particular forms of community and association with particular localities in a rather deterministic way. While Tonnies believed that Gemeinschaft type relationships involving kin and neighbours were more likely to be found in small-scale communities, he at the same time believed that friendship flourished in urban settings. Thus, while Tonnies praised rural life for manifesting close primary relationships that were affirmative of shared identity, Tonnies seemed to regard both place *and* type of relationship as necessary to understand the nature of bonds in any geographic community. Yet, Lee and Newby argue that since Tonnies wrote his seminal work more attention has been paid to the locality than to the type of relationship which could characterise the whole of society, a phenomenon they term 'communion' (Lee and Newby, 1983: 45). Lee and Newby have suggested that the problem has been that three elements commonly used to define community, community as locality, community as a local social system and community as a type of relationship have tended to be run together leading to confusion between empirical and normative conceptualisations of community Once these aspects of community were considered separately and empirically the rural–urban continuum came under increasing challenge (Lee and Newby, 1983: 57–58).

Calhoun rejects the idea that community belongs exclusively to either a geographic or cultural space, rather community is seen in terms of the quality of social relations. Thus Calhoun sees community as a pattern of social organisation and as a culturally defined way of life (Calhoun, 1980: 111). He suggests that there are three orders of communal bonds: those based on familiarity, specific obligations and diffuse obligations (Calhoun, 1980: 117). Calhoun argues that too much emphasis on the experiential aspects of community life understates the significance of the social bonds and political mechanisms which hold communities together or indeed cause divisions. The experiential dimension is not independent of the structural and community cannot be defined in terms of members' sense of belonging:

> What is important about 'sense of belonging' is not someone's identification of membership in a bounded collectivity, but his modification of his consideration of alternative courses of action on the basis of the communal relations to which he belongs (Calhoun, 1980: 110).

Calhoun rejects the idea that symbolic communities can offer support and belonging in the same way that a social grouping engaged in regular face-to-face interaction can:

> The communal may be regarded as a specialized sub-set of the social. In the connotations of everyday usage, community suggests a greater 'closeness' of relations than does society. This closeness seems to imply, though not rigidly, face to face contact, commonalty of purpose, familiarity and dependability (Calhoun, 1980: 111).

Differentiation within communities

Elias identified the idea of interdependencies as central to original understandings of the term 'community'. Elias points out that these interdependencies arise out of living in close proximity within visible and invisible boundaries. Furthermore, they are unevenly reciprocal as they involve power dimensions. However, as societies become more differentiated the bonds that connect people tend to be confined more to their private lives than to public life (Elias, 1974: xxix). In his research study *The Established and the Outsiders* Elias found that social class, gender and sociological oldness were all related to how connected people were in their private lives in a particular locality (Elias and Scotson, 1965). Members of the working class, women and old families were more likely to be bonded by community ties than members of the middle class, men or more recent residents. Elias explained class differences in sociality patterns by the possibility of sociability and leisure needs being satisfied by transport mobility and financial resources.

Social class

Class appears be a significant element in determining the extent to which we form significant relationships and commitments within and outside the localities where we live and patterns of neighbouring itself. Some writers, such as Gans, Pahl and Castells have argued that class is a decisive influence on the nature of relationships within localities. In the first instance, as Gans points out, the grouping of individuals who come together in a particular locality is determined by the outcome of a series of constrained choices on where to live and this is largely controlled by the housing market (Gans, 1962). Furthermore, he argued that subjective perceptions of others and objective material constraints, both connected with social class, could clearly be expected to influence relationships and lifestyle.

Pahl, writing in the 1960s about urban communities, and later Castells, drew attention to the ways in which national and global economic forces acting on localities frequently lead to conflict and marginalisation of less powerful groups. Both Pahl and Castells were concerned about the influence of powerful structural forces on local areas and about the conflict that arises between the large-scale and the small-scale.

Enforced communality

Enforced communality may result from poverty and inequality associated with class divisions. It has been found that in both rural and urban communities the lives of people from working-class or low-income backgrounds tend to be conducted within more constrained spatial boundaries while middle class and affluent groups are able to escape from the constraints of locality when they choose (Bartley, 1999: 242–243). Therefore, the degree to which networks of social relations are locality based is often a function of economic constraint. Having a car and the money to join say a golf club may be one aspect of lifestyle that differentiates older people's leisure activities and is clearly class related. But enforced communality can also be imposed in other ways. The power of state welfare bureaucracies to stereotype individuals should not be underestimated. We will discuss in Chapter Three how social policies may homogenise older people. While individuals may try to resist conformity to practices and values that are not compatible with their own self-identity, older people may find it difficult to exercise choices especially if they are in the Fourth Age.

Affluence appears to be associated with a greater desire for privacy and a greater sense of social distance in neighbourhoods (Bulmer, 1986: 85). Based on a study of six localities in England, Abrams found that middle class residents tended to have a different role and different relationships to their neighbour. Their range of contacts tended to be wider, their friends and family were more likely to be geographically distant, they were more likely to be actively involved in voluntary associations, and relations with neighbours were more likely to be on a formal basis. However, Abrams argued that class on its own was not sufficient to explain the extent and intensity of neighbourly relations, rather 'it needs to be related to wider trends towards privatisation, to dominant social norms, and to the nature of support networks' (Bulmer, 1986: 85).

Abrams expressed dissatisfaction with explanations that emphasise choice in middle class life and constraint in working class life (Bulmer, 1986: 40). Abrams argues that the stress on choice obscures the extent to which choice is linked to power and to constraint. Working class patterns of sociability are also chosen. The difference is that they are chosen on the basis of different resources and in the face of a different order of costs and opportunities. Therefore, class not only permits or denies choice but also shapes and structures it.

Social networks and personal communities

The concept of social network draws attention to the entire web of relationships, proximate and distant, emotional and instrumental, that a person accumulates over the life course: 'Rather than just focusing on one type of tie – say kinship, or neighbouring, or friendship – it provides a framework for appreciating the interplay of all these different relationships' (Allan, 1996: 126–7). Social network approaches have been widely used, particularly in Britain, to examine the set of informal ties in which people are embedded, and to investigate the resources older adults can draw on – particularly if they experience difficulties in meeting their day-to-day needs – by suggesting that older people have a potentially diverse range of relationships, comprising friends, neighbours, present or former workmates and kin (Phillipson *et al.*, 2001: 20, 26).

With the development of mass transport and communications systems there was perceived to be reduced significance or changed meaning of the 'local' in most people's lives. The idea of 'personal communities' has been advanced in opposition to the idea of place based communities to explain the extent to which significant relationships are often geographically dispersed but accessible through modern methods of communication. Wellman's work in Toronto, Canada was particularly influential in using social network approaches to understand community and the social supports that individuals could draw on for different purposes. On the basis of both qualitative research and survey research he conceptualised a move from kinship based dense local networks of relationships to 'personal communities' where each person's exchange and support network consisting of friends, neighbours, leisure associates and kin, was unique to themselves and was not geographically bounded (Wellman, 1990).

Phillipson *et al.* incorporated a spatial aspect into social network analysis in their comparative study of ageing in three different types of urban communities in London and surrounding cities. The authors found that different types of family and neighbour networks were associated with each locality as a result of different experiences of urban change. One locality was characterised as 'network poor' as kin and neighbour ties weakened due in part to particular migratory patterns while a second locality which had less drastic population changes was said to exhibit a pattern of 'local extended family'. In the third locality a pattern of 'dispersed extended family' was discerned (Phillipson *et al.*, 2001: 25–256). The advantage of such an approach is explained by the authors:

> The value of taking a locality perspective (as opposed to a national sample), is that we are able to view old age from a special vantage point, exploring the way in which support is produced through interactions in areas with specific histories and identities; hence the importance of place (Phillipson *et al.*, 2001: 37).

The contribution of the work by Phillipson *et al.* is that it shows that kinship and community are experienced in different and complex ways and the importance of understanding the interplay between individual biographies, opportunities in the structure of societies, urban change and cultural expectations.

A limitation of social network analysis is that it cannot fully answer questions about the sense of obligation and commitment we feel towards particular ties, nor the importance of non-instrumental and non-intimate relationships in maintaining a sense of self and of solidarity with others. Therefore, a focus on networks can lead to an over emphasis on observable roles and overt exchanges and can decontextualise the cultural aspects of mutual obligations. Crow suggests that we should not reduce relationships between people in networks to the properties of individual agents themselves; rather the challenge is 'to identify what it is about how people come together that enables them collectively to make the most of their individual skills and attributes' (Crow, 2004: 16).

Conceptualising community in the study design

We did take account of social network perspectives in both the ethnographic design of this study and in drawing up our questionnaire. We asked about a diverse range of relationships including those with family members, relatives, friends and neighbours and we tried to capture the variety of social worlds of our respondents. Examples of dispersed social networks that were important for older people were children and relatives living some distance away and Active Retirement groups attached to places of previous employment through which contact was maintained with former workmates. However, a more detailed examination is required of the specific interdependencies and exchanges that characterise localities in Rathmore and Rathbeg.

Thus further investigation of the concepts of neighbouring and reciprocities is called for.

Calhoun's analysis of community as both a pattern of social organisation and a culturally defined way of life emphasises that locality should be thought of in terms of multiplex social networks bounded by every day routines and interactions. Using both ethnographic and survey approaches we propose to map the significant social groupings in Rathmore and Rathbeg, to delineate the relationships of sociability and support involving family, kin, friends and neighbours, and elucidate the meaning and value that people derive from their interactions and relationships. Thus the specific interdependencies and the moral frameworks involved in older people's communities will be elucidated.

The significance of community as involving both place and type of relationship may be that it suggests the possibility that communities involving multiplex relations of family, kin and friendship sustained in face-to-face relations constitute a form of human life particularly well worth investigating. And while changes in society may have made this type of community less common, perhaps the search for authenticity that is said to be a feature of the present historical period may in fact demand multidimensional relationships of a similar type to those described by Tonnies.

Chapter Three
Ageing and Social Policies in Ireland

Introduction

In this chapter we examine the broad features of social policy towards older people in Ireland. Our intention is to explore how older people are viewed in public policies and to examine the solidarities and supports within communities that enhance personal and social well-being. We discuss the evolution of social policies and we examine the principles of intervention through welfare, support and care policies referring to the main policy documents relating to the older person. We ask to what extent public policies enable older people to live satisfying and integrated lives in conditions that are acceptable to them. We also address the practical effects of specific policies for different groups of older people in Rathmore and Rathbeg.

We begin by considering the ways in which social policies may categorise and stereotype older people and thereby deny them opportunities for full social and cultural integration. We give an overview of social policy provision for older people in Ireland, highlighting core services, the role of the voluntary sector and deficiencies in relation to community support services. The evolution of social policies for older people is traced through our examination of key reports and our discussion of principles such as healthy ageing, quality of life and ageing in place. We briefly review satisfaction levels with a number of social services among our study participants and we highlight the perception that more supports should be available. We focus in particular on a number of important community care services including the home help service and day care. We describe a number of day centres that provide opportunities for meaningful social interaction and vital involvement. Finally we examine the role of the voluntary sector in service provision in Rathbeg and Rathmore.

Public policies are not neutral but contain political values. Both policy and the absence of policy imply political positions. Social policies that affect the distribution of resources between groups not only create or exclude practical possibilities but they also create discourses that help to shape ideas and social perceptions or as Edmondson puts it 'worlds of expectation and justification' (Edmondson, 2000: 8). While policy makers may often present their proposals as factual and information based and as reflecting current standards and expectations, they implicitly create, support, or exclude certain values. For example, while private nursing home care is represented as providing choice to older people and their families and as part of a mixed economy of welfare for meeting long-term care needs, it can also be argued that the reliance on private nursing homes to provide care for over one half of all older people in need of long-term care in Ireland[1] suggests that caring for incapacitated and dependent older people is primarily a private affair between nursing home proprietor and individual older people and their families and is not a societal concern. The question of financing long-term care for dependent older people became embroiled in technical and legal controversy in Ireland in recent years.[2] Yet, the more complex and value-laden political questions raised, such as the moral and financial obligations of adult children towards their parents, are often inadequately debated.

Planning to meet the health and social needs of an expanding older population involves quantitative analysis that may classify older people as pension recipients, occupiers of long-stay beds and users of social care services such as home help or day centres – all competing

1 Private nursing home beds represented about 55 per cent of all long-stay beds in Ireland in 2001 (Department of Health and Children. Unpublished data).

2 In April 05 the Supreme Court ruled that charges for maintenance in long-stay hospitals and nursing homes were illegal and many older long-term residents and patients became entitled to a refund. In December 06 a new scheme for assessing and charging for long-term care in Ireland was announced which is due to come into effect in January 08.

with other budgetary priorities from public expenditure. However, the reality is that only a relatively small amount of care is given by social services – on average in the European Union 13 per cent of care is given by social services and three per cent by voluntary organisations (Eurobarometer report, 1993). By far the greatest proportion of care in older age is self-care. Other sources of care include spouse, children, other relatives, neighbours and friends.

Categorising for purposes of meeting care needs can lead to stereotyping, and in a society where being productive in the work force appears to be increasingly valued, older people, particularly those who are more 'dependent' on state services, may be perceived as a category of welfare that sidelines them into the margins of the real business of living. For example in the debate about difficulties currently being experienced in the hospital services in Ireland, sick or incapacitated older people have been represented as 'bed blockers' – a term that is particularly ageist. Furthermore, it is only in recent years that policy documents and research reports have based their recommendations on direct consultations with older people themselves.

Edmondson has argued that while older people may have status as a good cause she questions whether public policy enhances their social influence, social power or creates possibilities for them to achieve an excellent old age (Edmondson, 2000: 2). She suggests at least two reasons for this: firstly, the absence of input from older people themselves into policy making, and secondly, the inadequacy of the language available in the discourses used to express desirable outcomes and goals in policies for older people. It should be said that organisations representing older people are seeking to change perceptions of older people as passive and marginal. (A brief description of the main organisations representing older people can be found at Appendix B). Furthermore, the changing educational profile of the older population is likely to lead to higher expectations on the part of older people themselves for a more participative role in society.

Overview of social policy provision for older people in Ireland

Core services

Services for older people in Ireland are provided through a mixed economy of welfare model. While there are a number of important universal social services, for example free travel for every adult over 66 and free primary health care for everyone over 70 years, social services for older people as for other social groups have been delivered mainly on a selective basis. Core social services for older people include domiciliary care, public housing, subvention for nursing home care and non-contributory old age pensions. Theoretically these are provided on the basis of assessed need but the reality may leave much more to chance as the delivery of domiciliary services in particular does not occur in an orderly fashion (see for example Lundstrom and Mc Keown, 1994, Garavan *et al.*, 2001).

Community care

O'Shea has summarised the shortcomings of community care services for older people in Ireland (O'Shea, 2006). Pointing out that community care services are largely concentrated on public health nursing and on home help provision, he criticises the absence of a legal basis for the majority of community care services and the thin spread of services. For example meals on wheels are provided almost entirely by voluntary groups to less than 1 per cent of the elderly population, there is an acute shortage of paramedical and therapeutic services, and there are very few day care places available across the country (O'Shea, 2006: 16). Furthermore, O'Shea argues that 'community care services tend to accommodate the needs of providers rather than consumers with respect to opening hours, times of delivery and general availability' (O'Shea, 2006: 17).

It has been argued that the haphazard delivery of community care services and the legislative right to a subvention have acted as a perverse incentive for incapacitated older people to go into private nursing homes, contrary to government policy documents that have advocated 'ageing in place' with institutionalisation as a last resort (O'Shea, 2002: 118–119). The services and supports necessary to give effect to principles of 'ageing in place' have been slow to develop. Specific policy areas that remain underdeveloped include the home help service, rural public transport[3] and sheltered housing for semi-dependent older people (Layte *et al.*, 1999: 22). For example a study of the use of health and social services by older adults in the community found that among a sample of 937 older people in two health board areas in the country a substantial proportion, 37 per cent, of those found to be severely impaired in carrying out activities of daily living had not received any home-based services in the previous year (Garavan *et al.*, 2001: 23). It is generally agreed that we are starting from a very low support base in Ireland compared with other countries in Europe in relation to both domiciliary supports, day care services and community based medical services such as physiotherapists and occupational therapists.

Income support

Turning to income support it has been suggested that social welfare pensions in Ireland provide an effective and efficient policy instrument to deal with poverty in old age and one that people find acceptable. The flat-rate structure, the broad coverage of the social welfare pension system and the small gap between contributory and non-contributory pension benefits provide, according to the authors of a research report on well-being of older people, 'something approaching a state provided minimum income guarantee to older people' (Layte *et al.*, 1999: 21). Furthermore, the availability of non-cash benefits such as fuel allowance, free television licence and subsidised phone and

3 The Minister for Transport announced in April 05 that the Rural Transport Initiative then operating on a pilot basis would be put on a permanent footing.

electricity use increases the disposable income of elderly households. However, older women have cause for dissatisfaction with the qualifying conditions for the contributory old age pension.[4]

While elderly households are more likely to be found at the bottom of the income range they are not among the 10 per cent at the very bottom (a high proportion are non-elderly households with children) and the level of non-cash benefits is sufficient to lift a large proportion of the elderly households above the 50 per cent relative poverty line though not above the 60 per cent poverty line (Layte *et al.*, 1999: 54). However, it is private occupational pensions that make the difference between a modest life style on a basic income and a more comfortable lifestyle on an income proportionate to pre-retirement income. The proportion of occupational pensions among older people in Ireland is increasing (O'Shea, 1993). However, occupational pensions have always been class and gender related and are a mechanism whereby inequalities in life chances are carried across the life course. While private occupational pensions are promoted by public policy[5] to top up basic social welfare pensions, the fragmentation of the pension economy and the workforce, in addition to fluctuating returns from pension investment funds, has made the private pension market spectacularly volatile in recent years.

Role of voluntary sector

In the *Care of the Aged* report, 1968, the voluntary sector was identified as having a key role to play in providing social support services

4 At the time of the study the vast majority of homemakers in Ireland who historically at any rate are women had no social insurance and did not qualify for a contributory pension in their own right. Under a Homemaker's Scheme introduced in 1994 a period of time, up to a maximum of twenty years, may be disregarded at the pension decision stage where a person has registered with the relevant Department as a homemaker and is caring for a child under 12 years or incapacitated relative.

5 The Pension (Amendment) Act, 2002, introduced a new form of pension product, the Personal Retirement Savings Account which involves tax relief and is portable across jobs.

in partnership with families and the state, particularly in personal care services and organising social activities including home visiting. Voluntary organisations on behalf of older people grew significantly in the 1970s, facilitated by grants provided by the health boards for the establishment of Social Service Councils and Care of the Aged committees. Activities of voluntary organisations can range from providing a continuum of supported care to arranging an annual outing. The *Years Ahead* report, 1988 calculated the contribution of the voluntary sector as follows:

> As much as half of the home help service, almost all of the meals-on-wheels and laundry service and a sizeable proportion of day care centres for the elderly are run by voluntary organisations (The *Years Ahead*: 25).

While The *Years Ahead* was very positive about the role of the voluntary sector (The *Years Ahead*: 25), the report recognised that the voluntary sector is diverse, complex and fragmented and pointed out that not all voluntary activity is innovative, flexible and cost effective (The *Years Ahead*: 169–170). While a role was envisaged for the voluntary sector in local co-ordinating structures for service development and delivery, the community care teams within each Health Board took on many of the co-ordinating functions envisaged by the Care of the Aged report for the Social Service Councils. Furthermore, different models of voluntary involvement appear to have evolved in different parts of the country. We found that the level of cooperation between voluntary groups and public bodies and the capacity of voluntary groups to be innovative varied enormously. In Donegal there appeared to be very close cooperation between the voluntary sector and the North Western Health Board. A less effective partnership between some voluntary providers and the public bodies was noted in the Eastern Region. This is discussed in more detail at the end of the chapter.

Evolution of social policies

We begin our detailed examination of social policies by reviewing four key policy documents that have directly shaped policy and services for older people in Ireland:

1. The *Care of the Aged* report, 1968 which was the seminal report which addressed the needs of older people as a distinct group in a coherent way and dominated policy for twenty years,
2. *The Years Ahead – A Policy for the Elderly*, 1988 which is the basis of official policy for older people in Ireland,
3. *Shaping a Healthier Future: A Strategy for Effective Health Care in the 1990s*, which while concerned with the overall shape of the health services had particular relevance for services for ill and dependent older people, and
4. The 2001 Health Strategy, *Quality and Fairness – A Health System For You* which aims to provide improved social gain and quality of life for vulnerable groups including older people.

Care of the Aged *report*

The *Care of the Aged* report in 1968 was the first public policy review of welfare services for older people since the foundation of the State. Services for older people in Ireland in the 1960s were confined to a core number of income maintenance payments, most importantly old age pensions and a rather stigmatised system of residential care in County Homes, formerly Work Houses, for the infirm or chronically ill, usually poor older persons. This reflected the historical development of social services for older people that were rooted in the Poor Law and the low levels of investment in social services in Ireland since Independence, compared with other European countries.[6] The report recognised that services lagged behind those in other countries (*Care of the Aged*: 49) and it set out two principles to guide development of provisions for older people:

a) to enable the elderly person to live at home, where possible, at an optimum level of health and independence; and
b) to enable those who cannot live at home to receive treatment, rehabilitation and care in accommodation and in an environment as near as possible to home (*Care of the Aged*: ix).

These aims were to be achieved by an integrated range of services including welfare, health, social care and housing and appropriate residential care but based on a model of complementary care between the public authorities and the family (*Care of the Aged*: 13).

The philosophy of the report was one of encouraging a humane approach to older people experiencing incapacities of one kind or another, involving the development of community based social services, more caring attitudes among the public and encouraging the unpaid work of voluntary groups. However, while there is a strong sense of the dignity of each individual older person in the report, there is little sense of older people having a contribution to make to society. From the outset the report adopted a social policy approach that problematised old age as a 'care' issue and looked to improved social services to provide solutions. It also envisaged a continuing role for the voluntary sector and commended a practice of voluntary groups and schoolchildren visiting older people.

While this approach was well-meaning, a more positive vision of ageing would recast the terms of such relationships from one-sided giving and receiving to emphasising the mutual advantages from such friendships. An independent study of the implementation of the 1988 *Years Ahead* report carried out for the National Council on Ageing and Older People proposed such a vision of ageing in which

6 Fanning (1999: 55) explains that the role of the State as provider within the Irish mixed economy of welfare became less developed than in Britain due to resistance to collectivization within Irish society and for reasons of affordability.

> Older people engage in reciprocal and mutually beneficial relationships with people in other age groups rather than being seen solely as 'receivers of care' from others who are more resourceful and able (Ruddle *et al.*, 1997: 326).

The Years Ahead: A Policy for the Elderly

The *Years Ahead* report, published 20 years after the *Care of the Aged* report and constituting official policy for the development of services for older people in Ireland, restated the principles of the earlier report in relation to supporting a good quality of life at home in conjunction with family, professionals and community and providing high quality alternative care where that was not possible (The *Years Ahead*: 38). The *Years Ahead* advocated a strong service delivery model. There were over 120 main recommendations relating to health, housing and social care services for older people at home, in the community, in hospitals and in long-term care. The role of key health professionals and social care providers was emphasised. This focus reflects the growth in professional health and welfare services between the publication of the two reports and in particular following the establishment of community care teams under the new health boards set up in 1972.

The *Years Ahead* articulated a positive vision of old age and beliefs about what constituted a good quality of life:

> The health and welfare of the elderly is greatly affected by whether or not they consider themselves to be valued members of the community … the years after 65 can often afford the individual an ideal opportunity for increased personal fulfilment and time to enrich the community with the contribution of accumulated wisdom, life and practical skills. It is, therefore, important to make every effort to develop a social, cultural and economic milieu in which older people can participate and contribute effectively (The *Years Ahead*, 1988: 61).

Subsequent policy reports, including those with a 'healthy ageing' focus and those concerned with quality of life for older people in need of long-term care, have continued to articulate a positive vision of the social and cultural integration of older people in their communities.

In addition to social care supports for ageing in place, The *Years Ahead* recommended that a wider range of facilities be developed to meet both the low to medium and the high dependency needs of frail older people. Community hospitals were to provide a range of services, such as assessment and rehabilitation, high dependency nursing, convalescent care, information and support for carers and day care (The *Years Ahead*: 134), while sheltered housing with back-up day care facilities and multi-purpose homes (as had been developed in Donegal) were to provide for the less dependent older person (The *Years Ahead*: 142). The *Years Ahead* also recognised the role that private nursing homes play in caring for older people and in the right of older people to avail of such care as a matter of choice (The *Years Ahead*: 142).

Shaping a Healthier Future

The emphasis on professional expertise as the key factor in responding to the health and social care needs of groups such as older people was challenged in the 1994 Health Strategy – *Shaping a Healthier Future* published by the Department of Health. There are two new stances implicit in the 1994 document: firstly, one cannot assume that services have an inherent value, it is necessary to evaluate the health and social gain they produce; secondly, it is essential to have consumer participation in the planning of services and accountability of service providers. There is evidence that efforts have been made by the health authorities to consult service users and their carers in reviewing and developing services.[7]

7 The Eastern Health Board consulted with service users in preparing the *Ten Year Action Plan for Services for Older People*, 1999 and *Review of the Action Plan*, 2001. A major evaluation of health and social services for older people, Garavan *et al.*, 2001, was based on the experiences and views of older people.

A refocus on services to concentrate on 'healthy ageing' strategies indicates a broader approach to policy development since the late 1980s. Healthy ageing implies a focus on the maintenance of health through lifestyle choices and preventive measures but it also includes the idea of 'active ageing' which the World Health Organisation defined as: 'the process of optimising opportunities for physical, social and mental well-being throughout the life course, in order to extend healthy life expectancy, productivity and quality of life in older age' (WHO, *Active Ageing*, 2002)

A Health Promotion Strategy, *Adding Years to Life and Life to Years*, was produced for older people in 1998. Health promotion strategies are now in place in all health boards in the country. Many are designed to promote physically active lifestyles among older people and to promote opportunities for social interaction. An indication of the broad scope of Healthy Ageing projects and of the range of factors that contribute to quality of life is seen in the description of ten projects from a Healthy Ageing Database of over 300 projects developed by the National Council on Ageing and Older People (O'Shea, 2003). The projects which are based in a variety of settings including day centres, clubs, extended care settings, community settings, or older persons' own homes, cover the following five policy areas: building healthy public policy, creating supportive environments, strengthening community action, developing personal skills and reorienting health services.[8]

Health promotion programmes are in the main one-off, low-budget pilot programmes limited to specific catchment areas or specific target groups of older people and are often seen as a luxury that can be dispensed with when cutbacks are made in mainstream health

---

8 Examples of programmes include 'Go for Life', a national joint initiative programme by Age and Opportunity and the Irish Sports Council to promote physical exercise and sport activity to suit all levels of fitness, 'Voice of Older People' – a community development project in the North Western Health Board where older people tackle health and social issues that affect them, and 'Arts for Older People' in care settings in the Midlands Health Board.

and welfare services. While they may impact positively on some aspect of the lives of a small minority of older people and may help to promote positive attitudes to ageing, there is a danger that they will have no lasting impact unless they are designed to challenge and change existing institutional practices.

There is a danger therefore that 'new' health promotion programmes with an educational focus and an emphasis on 'lifestyle' alone could be seen as offering answers to complex questions about how to achieve greater social integration of older people. Clearly they should not deflect attention away from planning and resourcing longer-term policies that more directly affect social participation and cultural integration such as adequate income, housing, recreational and environmental facilities. For example in the evaluation study of Healthy Ageing projects in Ireland there was a poor representation of projects in the housing and transport sectors, which was surprising given the importance of both sectors in the lives of older people. The author suggests that part of the problem may be that both transport and housing require more integrated development than what has been achieved to date and that there has been an overemphasis on health-care services as the solution to all the problems associated with healthy ageing (O'Shea, 2003: 81). Furthermore, the well established links between socio-economic status and health (see for example Davey-Smith *et al.*, 1994) and the evidence that educational campaigns to promote healthier individual lifestyles appear to achieve more success among middle class groups (Collins and Shelley, 1997) suggest that we need to examine the effectiveness of standardised health promotion programmes among all social categories of older adults.

Quality and Fairness – A Health System For You

The 2001 Health Strategy highlighted quality of life as one of its central objectives and proposed a number of actions to develop a comprehensive approach to meeting the needs of older people into the future.[9]

---

9 There are specific actions proposed including: health promotion, dementia services, clarification on eligibility and subvention arrangements for long-term

While the 2001 Strategy shows an awareness of the many difficulties facing older people in Ireland and is more person-centred and holistic in outlook than previous policy documents, it is argued that the system remains medically-oriented with little attention paid to social models of ageing (O'Shea, 2006: 16). We consider policy approaches to improving quality of life in the next section.

Quality of life and social policies

Studies of quality of life in old age have become of considerable interest to policy makers in recent years.[10] In response to criticism that some studies have tended to develop rather static indicators of well-being that fail to capture the social context and multi-layered processes involved, recent studies have sought to conceptualise and measure quality of life from older people's understandings of what is important and meaningful for them. Bowling *et al.*, 2002 provide a multidimensional model of quality of life that captures individual and communal dimensions and Murphy *et al.*, 2006, have adopted a similar approach in the Irish context in relation to older people in extended care settings.

Quality of life in extended care facilities

Long-stay care for older people with dependencies is currently a particular focus of public policy in Ireland. Recent scandals have focused

care, provision of additional community nursing unit places, extension of the remit of the Social Services Inspectorate to include residential care for older people, and preparation of national standards for community and long-term residential care.

10 The focus of the ESRC Growing Older programme in the UK which ran from 2001 until 2004 was quality of life. This included large scale surveys that measured variation in quality of life to in-depth interviews and focus groups with particular categories of older people.

attention on quality issues in residential care. These have related to the reform of the subvention scheme for private nursing home care, eligibility criteria, quality controls and inspection. Only private and voluntary nursing homes have been regulated under the Health (Nursing Homes) Act 1990, with public facilities falling outside the legislation. There is general agreement that existing standards need to be updated and developed in line with current expectations and ideas about quality of care and quality of life. Identifying the relevant domains of quality of life for people living in residential care is critical to making progress in this area (O'Shea, 2006: 22). Draft National Standards for Residential Care Settings were published in January 07 by the Department of Health and Children. The proposed regulatory framework is based on the views of service providers and service users and was underpinned by research that identified quality of life domains for residents in long-stay care in Ireland. In a recent study Murphy *et al.* (2006) found that what residents valued included:

- Independence, privacy, and self expression,
- Confirmation that their life contains meaning and hope in the same measures as before,
- A homely atmosphere and person-centred routines that emphasise choice and empowerment,
- Aesthetically pleasing physical environment,
- To be connected to other people and to be able to maintain relationships with family and friends,
- To have meaningful and purposeful activity which they themselves are involved in planning and arranging.

The new draft standards include requirements in relation to social connectedness, social relationships, activities and community contact. The importance of these aspects of care is highlighted in the residential and day care facilities described later in the chapter.

Nonetheless, questions remain as to whether we have the correct balance between public and private provision, the location of long-stay facilities and what we should expect from them in terms of the integration of residents with local social networks and community:

> Apart from the failure to develop a fair and transparent eligibility and financing system for extended care, there are a number of other aspects of public policy that of concern: firstly, the failure to develop models and understandings of what constitutes a good care environment and secondly, the failure to plan for residential care facilities that are located as close as possible to people's communities and that provide a wide variety of facilities that maximize opportunities for social interaction and normalised routines (Gallagher, 2006: 66–67).

Many private nursing homes appear to be extremely isolated from the surrounding locality and from community networks such as social clubs and parish groups (Gallagher, 2006: 70). A model of public provision of residential and day care that tries to promote both creativity and connectedness can be seen in the Community Units built in recent years by the health boards:

> The model of the small purpose built Community Unit that includes day care and has dedicated activities staff appears to provide the possibility for developing a more normal, lively and more community oriented facility ... They are less isolated from communities, they are resourced to provide more individualised programmes of animation, there is a wider range of developmental and recreational activities on offer and there is a sense of serving and being part of the community (Gallagher, 2005).

We will give examples of good models of locally based provision for older people that seek to provide good standards of care and to integrate dependent older people in the surrounding community.

General satisfaction levels with social services

Many respondents in the survey commented favourably on social services provision for older people, highlighting in particular free travel and the medical card (free primary health care) as services they valued. One respondent stated: 'We are well looked after even by the Government at the moment, for example the free travel. We have a lot to be thankful for compared to England.' In general we found that respondents who were availing of community care services were satisfied with these including the home help service, day care services and the meals service. Satisfaction levels with the health services were also generally high. Respondents reported themselves broadly satisfied with the general practitioner service and public nursing service, however, there was a perception that there was a shortage of public health nurses and transport to/from hospital was criticised by a few respondents. One of the key informants was very critical of the health services, in particular the hospital services. This was based on unsatisfactory experiences of the care given to her invalided husband during many episodes when he had to be admitted to hospital from home where she cared for him on a full-time basis.

Issues for residents of sheltered housing

Sheltered housing units are allocated by the local authorities to older people under the scheme of priority for letting. Each single unit is approximately 180 sq ft in area and consists of a bed-sitting room, bathroom, small kitchenette and hallway. Each complex or court has a community room for meals and communal social activities and a laundry room. Applicants for sheltered housing used to have to be at least 60 years. However, the age limit was reduced to 55 which has brought a more mixed age population into the complexes. Some of the residents have expressed misgivings about this. Dublin City Council officials themselves acknowledged that in some complexes there are many people with alcohol problems and that their vetting system may not have been tight enough.

Sheltered Housing Liaison Officers (formerly known as Wardens) have responsibility for surveillance and support of residents. They check that residents are all right, deal with problems in relation to their accommodation, coordinate support from the community health services and liaise with Residents' Committees. The job of the Liaison Officer was expanded and developed during 2002–2003. They were given a role in the allocation of units, a central administrative office, more back-up and support through written guidelines and flexibility in rotas and a greater role in promoting and coordinating social and recreational activities in their courts. The community rooms have remained a rather underutilised resource. Most complexes had a meals service and Mass generally took place once a month. However, while some complexes had several regular activities such as art and bridge, other complexes had no organised social activities at all. During the period of the study the City Council had begun to provide more funding for classes and social events such as painting, dancing and summer community festivals.

Dublin City Council are cooperating with the health authorities in developing more integrated models of care. This involves for example providing health and day care facilities in conjunction with sheltered housing but also serving the surrounding community. It appears that the Council wishes to adopt a less paternalistic role while at the same time providing appropriate support services and opportunities for social interaction. This may help to remove a certain negative label associated with sheltered housing. The Council are increasingly referring to the complexes as 'supported living' arguing that many residents need little back-up.

The difficulty of striking the correct balance in terms of support was underlined in the contrasting perceptions we found among residents of the purpose of sheltered housing and the role of the Liaison Officers. Some residents regretted that there was not a resident warden while others, such as a resident who found the Liaison Officer 'intrusive', felt that the very role of a Liaison Officer institutionalised the complex and stigmatised the residents. These contrasting views appeared to be based on both the level of dependence of individual residents and the expectations of residents prior to moving in.

When we asked residents how satisfied they were with the service provided by the Liaison Officer of the 18 residents we interviewed 14 were satisfied with the service while 4 were not. Most residents valued the regular contact with the Liaison Officer and some of the residents were fulsome in their praise. One resident stated: '[name of Liaison Officer] is an excellent warden. When she heard I was sick she came straight up to see if I needed the doctor'. (Interview July 02). However, not everybody 'got on' with the Liaison Officer and there were differences in personalities and conflicting expectations that resulted in a small number of residents expressing criticism including one or two who cited favouritism. As already stated one resident believed that the Liaison Officer was intrusive, however, her concern was more with the role of the Liaison Officer than the personality of the incumbent.

A major cause of dissatisfaction among the residents was the size of the accommodation itself. A significant number (6 out of 18) of the residents interviewed in the survey expressed dissatisfaction with their rather cramped accommodation, especially having a bed-sitting room rather than a separate living room and separate bedroom. The author who carried out the interviews herself in the residents' apartments was able to observe at first hand the lack of privacy involved.

Need for more supports and involvement

However, while satisfaction levels were reasonably high in relation to domiciliary care and health services, such supports may not meet needs for 'vital involvement' and social interaction. Therefore while respondents may have been satisfied with services, it was apparent from their replies to other questions that they believed that other types of supports and solidarities could be better developed such as 'befriending' schemes. Many respondents said that more could be done in localities for older people, especially for those living alone, such as visiting and doing small chores for them. Some suggested that younger people should do more for older people who lived alone and were not robust: 'Younger people should step in the shoes, help the people who have no one, people should come in and do some gardening and try to make their lives a little easier'.

The idea of secondary school students visiting older people they know was also proposed. However, the idea of older people having something to contribute was strongly suggested in the proposal made by three or four respondents that older people could help younger people. Suggestions made included teaching them skills they had such as cooking and sewing, talking to them about the past, helping them to appreciate the difficulties that life can bring and helping those who had a drugs problem. Many also believed that more opportunities should be provided for social interaction through clubs and community centres. One voluntary worker in Rathmore remarked: 'Older people are looked after materially and individually but not communally' (Interview with voluntary worker, 19/11/02). It is the communal aspect of support that we are particularly interested in; therefore the focus in subsequent chapters will not be on state health and welfare services as such but on communal provision through clubs and day centres that involves voluntary activity, on the caring networks found in communities and on the opportunities for older people themselves to participate and to contribute within their localities.

Respondents' experiences of community care services

Since our primary interest is in investigating the solidarities and supports within communities that enhance personal and social well-being we will examine in more detail the experiences of our respondents in relation to community support services. We focus on two services in particular, the home help service and day centres that are of particular importance in enabling older people to continue to live at home when they experience some degree of incapacity.

Home Help service

The Irish home help service is discretionary and has been criticised for the patchy and somewhat random distribution of home help provision for older people (Garavan *et al.*, 2001, Lundstrom and Mc Keown, 1994). On the other hand those who do receive the service are generally very happy with it. Lundstrom summed up the service as follows: 'We can say of the home help service in the Republic of Ireland that it may be difficult to get, but when you get it, you get it good!' (Lundstrom 1995). Within the Eastern Region Health authority it is estimated that only 3.5 per cent of over 65s receive a home help service (EHB, 1999) while the figure is just under 5 per cent nationally (NCAOP, 1999). While underprovision is a problem, the strength of the Irish home help service may be its flexibility and its local base. This means that a home help can be provided who is sensitive to different situations and needs and can offer a mix of practical help, a watch and ward function, befriending and/or providing company. This can enable an older people to continue to live at home with normal routines, in familiar surroundings and on the basis of reciprocities.

A recent study of loneliness and social isolation among older Irish people indicated that the home help service was of major importance to the participants in the study with many regarding their home help as their special friend rather than as paid staff from an organisation: 'The flexibility and responsiveness of individual home helps were key to their great success in promoting the social networks of the participants' (Treacy *et al.*, 2005: 112). At the same time there is a serious problem of underprovision, the service needs to provide reliability and consistency and to operate to professional standards.

Of the 15 respondents in the present study (9.1 per cent of the sample) who had a home help, 13 were satisfied with their home help, one person was fairly satisfied and one was not satisfied. One informant in Rathbeg said that the home helps were great, they did over and above what their official duties required and took a real interest in their clients. On the other hand a voluntary worker in Rathbeg expressed the view that as the service had become more bureaucratised and professionalised the quality of care was not as good. An informant in Rathmore who is in a wheelchair spoke about how she would like her home help to do the hoovering but did not want to ask her: 'I wouldn't like to complain you know that kind of thing.' One informant in Rathmore who worked as a home help spoke about her flexible approach to her work:

> Now as I say I'm sent to a house and I'm supposed to be a carer and I'll do anything that needs to be done, including if there's dishes from the rest of the family, I'll do them. I don't believe in demarcation. I mean when I go out to Miss – tonight I'll take her curtains down and bring them back here and wash them. But really I don't have to do anything like that. But I do, because for one thing I have a pride in my job. I don't have any recognition as such but I know the people that I care for, if their families go abroad they never come back without a present for me (Interview 19/10/01).

The above selection of views illustrates that while the home help service is generally viewed in a very positive light, perceptions and experiences are also quite varied.

At the time of the study the management of the home help service represented a mix of voluntary and health board input and appeared to operate with few formalised guidelines from funding agencies. In the HSE Eastern region the service is provided via voluntary organisations while in the HSE North Western region it is provided directly by the HSE. While improvements in the rate of pay[11] have led to better recognition of the importance of the service that home helps provide, there are important issues related to retention of home helps,[12] supervision and training and the overall management of the service on the ground that urgently need to be addressed (NCAOP, 1999, O'Sullivan, 2003). If it is desirable, for example, to have input from the voluntary sector in the home help service then this needs to be properly planned for and supported through capacity building involving training and supervision and integration with other community care programmes. Given the sensitive nature of the work, the need for flexibility in work arrangements, the negotiated nature of familial care and the fact that clients do not always readily accept a home help, it is important that the service should seek to support and extend existing care networks.

Day care provision

Day care centres provide medical and social care services that are important in the continuum of care[13] while day centres and clubs for older people play an important role in activating social interaction.

---

11 The rate of pay for home helps is between €13.30 and €13.79 per hour (information from Home Help Organiser Rathmore 7/11/05).

12 The recent success of the Irish economy has provided alternative employment for many women who heretofore were attracted to this work.

13 Day care centres provide a fuller range of therapeutic and personal care services such as bathing, physiotherapy and nursing and are usually attached to public geriatric hospitals, community units, voluntary nursing homes and other long-stay facilities. Service users are more likely to have impaired mobility or cognitive difficulties and referrals are usually made by public health nurses.

Like the home help service, provision of day care is also a discretionary service and involves a mix of public and voluntary input and different models of service provision. Studies have pointed to a lack of consistency in the availability of day care throughout the country, shortage of places and a perception that day centres could achieve much more (Convery, 1987, Gallagher and Kennedy, 2003, Haslett, 2003).

While the number of day centres and clubs for older people has expanded in recent years,[14] there is a shortage of day care places, particularly in the Eastern Region. At the time of the study it was estimated that there were 2000 day care places in the Eastern Region but the reality is that such places are only available to individuals for one or two days a week (Review of the Implementation of the Ten Year Action Plan, 2001:11). The only day care centre in Rathmore had a waiting list of 80 persons[15] and most of those who do attend can only do so on one day per week. The unevenness of day care provision around the country is illustrated by the availability of five-day day care in one of the voluntary facilities near Rathbeg.

Day centres are mainly managed by locally based voluntary/parish groups and are grant aided by the health boards. Their focus is mainly recreational and they usually provide a meal. Recent research has found that while many social clubs/day centres run by voluntary organisations may have started out only having a social mandate, due to a chronic lack of day care centres many of these have ended up providing services to a wide spectrum of older people (Haslett, 2003).

Because of their social role and because interactions in day centres formed an important part of the ethnographic data gathered for this study, we pay particular attention to day centres in the study. The majority of day centres are operated by voluntary groups in local or parish centres and while they are grant aided by the health boards,

---

14 It was estimated that in 1982 in the Eastern Health Board there were 51 such facilities (The *Years Ahead*: 108) while in 1999 there were approximately 156 day centres or clubs for older people (Ten Year Action Plan: 37).

15 Information from Manager of Services for Older People, Community Care area 8 on 18/9/03.

they also rely on fundraising. The standard of facilities varies greatly[16] ranging from welcoming and homely to drab and dispiriting. In general we found a higher standard of finish and décor in the voluntary facilities than in health board facilities. We describe below the activities, atmosphere and general surroundings of day centre facilities in Rathmore and Rathbeg.

Involvement in sociable and purposive activities provides opportunities for creativity and connectedness that Moody suggested were so important for older adults and can be seen to represent approximately the type of 'vital involvement' that Erikson advocated. More attention has recently been paid to the quality of social and recreational activities in both day and residential settings. Recognising that older people have a capacity for continuing learning and activation of their creativity, many initiatives have been undertaken to provide meaningful activities programmes in care settings (Gallagher, 2005a). Activities such as drama, art, story telling, craft work and creative writing have been introduced successfully in day and residential settings and have challenged perceptions that passive leisure and entertainment such as bingo or sing-songs are sufficient to satisfy older people.

Observations in day centres

While we found contrasting approaches in day centres in relation to the extent to which service users were encouraged to do creative work themselves or learn new things or given opportunities to socialise with the wider community, it appeared that many day centres were open to trying out new approaches and utilising community resources such as libraries, schools and health promotion programmes. Two programmes 'Go for Life', a programme of physical exercise, and 'Arts

16 Standard accommodation in a day centre includes a kitchen, dining room and main day room for activities. Those centres that provide full day care facilities have a treatment room, shower room and laundry room. Some centres have a sun lounge or conservatory and a room for alternative use. Those attached to a long-stay facility generally have an Oratory.

in Care Settings' have been introduced successfully in many care settings.

From visits to and conversations with many day centre users it appeared that those who attended enjoyed the social interaction. Many enjoyed taking part in a specific activity and many commented that it was a welcome break from their everyday routine. Relationships and atmosphere between the staff and the participants themselves seemed invariably warm and relaxed. While some choice of activity was provided, we observed differences in regard to opportunities for creative and developmental activities, links with the wider community and degree of participation by older people themselves in the running of the centres.

For example we observed contrasting approaches to craft making. In one centre a group of women were making cards under the direction of an instructor and they spoke about the tremendous enjoyment they got from this activity. In another centre we observed that a staff member herself was making cards for sale for fundraising and the participants were not involved in this activity. Perhaps the problem was that the second centre had insufficient staffing and financial resources to have a Craft instructor or that the service users had significant disabilities. Nonetheless a vision of what is desirable in terms of purposive leisure has to be imagined by those who run such centres in order to facilitate the possibility of creative experiences for the service users.

It was interesting to note the varied disciplines from which supervisors of day centres are recruited and to reflect on the different approaches that may be associated with different backgrounds. For example, many supervisors are from nursing and care attendant backgrounds. However, in one particular day facility the supervisor was from a youth work background. She encouraged the users to decide what programmes and activities they wanted and to be involved in the running of the centre. This was interesting in terms of the discourses used in relation to older people; practices have tended to view them as passive and vulnerable whereas principles of empowerment were clearly in evidence here.

Visit to a day centre in Rathbeg

This day centre is located in a geographically dispersed rural area in the environs of Rathbeg. The design and physical appearance of the day centre is most attractive.[17] In operation since 1997 and developed by the local Care of the Aged committee in cooperation with public bodies, it was particularly well finished and decorated, looking more like a private house than an institution. Features and finishes such as a beautifully carved fireplace, coving and a subtle contrasting colour scheme were achieved as a result of the construction being undertaken as part of a FAS training scheme (FAS is the state training and employment agency). The voluntary management committee put a lot of effort into selecting good quality furniture and curtains. The reception area had a comfortable couch where you could sit on your own or have a quiet chat with someone. Paintings donated by a local artist hung on the dining room wall and plants were placed in a conservatory. The effect was very pleasing and comfortable. The attractiveness of this day centre was indicative of the vision, commitment and energy of the voluntary committee. Their vision articulated in their Mission statement emphasised the contribution that older people can make to their communities:

> We want to demonstrate and prove that by giving the elderly people in the community an opportunity to get together, their attitudes to life and living will be more positive and as a result they will remain healthier, and also hope that their values and experiences and skills might be passed on to the younger generation (Rathbeg Day Centre Committee-Development plan 2003).

This vision was put into practice through personal involvement in the centre from conception to completion. Good linkages were forged between state agencies (the health board and FAS) and within the community itself and this cooperation generated good will by the creation of training and employment opportunities locally. The committee were very involved in the day to day running of the centre and spoke about the activities in the centre, how important it was to pull in

17 The author visited this day centre on 18/8/03 and had an informal interview and tour of the facility with two volunteer members of the management committee.

resources from community arts, library facilities and health board programmes. They said they had to encourage the staff to keep up some programmes such as health promotion initiatives like 'Go for Life'. An interesting comment was that the service users coming to the centre dress up and the women put on jewellery. The committee saw this as positive, also reporting that the day centre users loved coming and many had made good friends there.

What difference if any does design and décor make? A social worker working with older people in the North Western health board[18] suggested that creating a homely atmosphere in a community hospital was not about design and furnishings and that it was a bit condescending to assume that all residents wished to bring their furniture in. In his experience what people want is a feeling of security and to be listened to. Indeed the day hospital near Rathbeg is a functional and plain facility with no pictures or home furnishings. Yet the open and warm approach of the staff towards the day users, including their willingness to consult them in developing interesting activities and to respond to their concerns, was evident in the range of activities provided and the sensitive way they looked after their personal and private needs. Lack of space to do alternative activities was the biggest design problem. For example the Oratory had to double as a place for the SONAS programme[19] Thus, while physical and design features are secondary to quality in caring relationships, attention to design and décor may reflect a vision of what a day centre should strive for and may suggest that older people are entitled to the highest standards of facilities that can be achieved.

Day care and social interaction

The provision of leisure/social activities for older people provides opportunities for social engagement by older people either as volunteers or as participants and day centres can be regarded as constructed

18 Interview with the social worker in the Community Hospital on 13/8/03.

19 SONAS is a multi-sensorial programme used in many residential settings for older people particularly those with sensory or cognitive impairment.

communities. Day centres are meeting places that may be especially important for those older people who experience isolation or loneliness. However, among our respondents only five were regular users of a day care centre and one went occasionally. While all expressed themselves satisfied with the service, many commenting on how they valued 'getting Mass' at one of the centres, day centres do not have universal appeal among older people. In particular older men appear reluctant to attend despite a wide range of social activities offered (Haslett, 2003, Davidson *et al.*, 2003).

While the overwhelming view that came across in Rathmore and Rathbeg was that social participation through involvement in clubs and groups that catered specifically for the older person was positive and life enhancing, some dissenting views were expressed about the desirability of older people socialising together. Two male respondents expressed negative feelings about Senior Citizens' clubs for example, one stating that they were 'depressing' and that it was 'better to keep out of them' while the other stated that he 'couldn't stand older people in bulk'. These men were by no means socially isolated. One of them, a 76 year old married man was particularly active – he played cards and snooker regularly and took part in dancing through the local GAA club. We will widen our discussion of social interaction in Chapter Four when we examine a broader range of social and leisure interests and activities.

The voluntary sector in Rathbeg

Care of the Aged committees were established in Donegal in the 1960s. There are forty-four Care of the Aged committees in the County, most but not all of which are active. The committees are cross-denominational and there is a coordinating committee that meets four or five times a year. Each committee receives a grant from the Health Board (now the HSE). The services and activities provided by the committees include laundry, chiropody, home security, help with accessing housing grants, outings and social functions. The *Years Ahead* commented that:

> Co-operation between the Care of the Aged Committees in Donegal and the North Western Health Board has been the foundation of much improved services for the elderly in that county' (The *Years Ahead*: 169).

As already stated their responsibilities have decreased as health boards have assumed direct responsibility for schemes that the voluntary committees administered in the past, for example the home help service. However some committees have interpreted their remit in a particularly ambitious and visionary way by organising themselves to become the vehicle for new sheltered housing and day care facilities.[20]

The high levels of cooperation between voluntary groups and the Health Board in Donegal was very evident in the comments made by both voluntary providers and Health Board staff and in the quality of services delivered. The Chairperson of the coordinating body for all the voluntary committees said that every committee had direct access to Health Board staff and he believed this made a big difference to the motivation and effectiveness of the committees. He pointed out that voluntary committees observe problems and needs locally and pass on information to the Public Health Nurse and Home Help Organiser. A voluntary committee member described the health board as 'very community oriented'. The Health Board's Service Manager for Older People in Donegal said that without the voluntary committees Donegal would have far fewer services. She said her staff met regularly with the committees and she praised their commitment. She also praised the work of St Vincent de Paul who do somewhat similar work.

20 Two purpose built facilities near Rathbeg were visited. They comprise a day centre and sheltered housing units while one has in addition a high support residential unit. Both complexes were built through cooperation between local voluntary committees and public bodies.

The voluntary sector in Rathmore

The voluntary organisations that are grant aided by the Health Board to provide services for older people in Community Care Area 8[21] within which Rathmore is located are: 17 organisations providing a meals service, 6 organisations providing home help services, 4 day care centres and 9 day centres.[22] During the period of the study the principal voluntary organisation that provides services for older people in Rathmore experienced difficulties and tensions. The centre which was already operating a home help service, a meals service, a day centre and a counselling service was required to expand its role in the provision of meals. However, new health and safety regulations and a desire by the public bodies to achieve more consistent standards, coordination and accountability led to the employment of more paid staff and a diminished role for volunteers who had provided the service for many years on an unpaid basis. The manner in which this change was handled led to deep levels of dissatisfaction among the volunteers.

Both the voluntary committee and the staff of the Health Board and the City Council seemed to share a view that the best service was not being delivered. The committee argued that in the first place they were not integrated sufficiently with the community care services run by the health board. They lacked the structures and resources to take on new responsibilities. Moreover they felt in a limbo, arguing that no clear policy direction was being given by the public bodies. The public bodies for their part appeared critical of the work of the voluntary committee itself, pointing out that they tended to be inflexible and somewhat negative about change. The committee did not seem to have attracted younger people or people with new ideas and skills.

While it is beyond the scope of the present study to analyse the reasons for poor coordination and cooperation between one voluntary organisation and the public bodies, some factors have suggested themselves. Many voluntary groups and statutory bodies report that there

---

21 Community Care area 8 has a total population of 193,000 of whom 7.3 per cent are 65 or older (Eastern Health Board Ten Year Action Plan: 105).

22 Information from Manager of Services for Older People, CCA 8.

are fewer people volunteering, therefore continuity and change can be difficult to accomplish. The Health Board do not appear to have been proactive in supporting and encouraging voluntary groups in their areas through training or information sessions. This is a wider problem that has been addressed in the White Paper on Supporting Voluntary Activity (2000). Voluntary groups find it frustrating that they get insufficient direction on policy changes. This may reflect the crises in health administration that appear to be particularly acute in the Eastern region in recent years.[23] The lack of capacity building in the Eastern region may be contrasted with the more effective integration of the voluntary Care of the Aged committees in Donegal. The success of the Care of the Aged committees seemed in part to depend on the Health Board working closely with the committees, encouraging coordination and keeping members informed of policy developments. Through regular seminars organised by the Health Board policy changes are explained and new ideas about active ageing for example are discussed. This appears to help the members to continue to respond in relevant ways to the social needs of older people and to merge their efforts with the public bodies.

Summary

Referring to our exploration of the idea of community itself in Chapter Two and to the possibility of multidimensional relationships we can see the challenges involved in providing good quality community supports and services for older people. The extent to which community implies obligation and makes demands, not just on the state as provider of welfare services, but on individuals themselves, families and citizens is often overlooked. It is sometimes assumed that the appellation 'community' somehow guarantees the provision of good stand-

23 The reorganisation of the Eastern region in 2000 into three area health boards and its more recent restructuring under the Health Services Executive may have absorbed much energy and left officials with little time for development work.

ards and acceptable forms of care. Yet it is clear from our examination that values, reciprocities, altruism and lifestyle choices are all involved in the experiences older people have of solidarity and support.

In this chapter we demonstrated that public policies towards older people imply a particular view of the possibilities of the latter stages of the life course and that some social policies have tended to categorise older people and marginalise them. We have examined changing perceptions of older people from needy, deserving dependents and passive recipients of care to autonomous adults who desire to maintain independence and to engage in normal social activities and relationships regardless of incapacities associated with ageing. We have examined policy ideas in relation to 'ageing in place', 'healthy ageing' and 'social gain' and the extent to which policy involves input from older people themselves. We discussed the evolution of social policies in terms of service development and the emergence of new concepts to express desirable goals and outcomes. The shortcomings of both institutional and community care provision were also examined. We described a number of community support services in some detail and we considered some survey evidence on the experiences of older people in relation to the supports provided. We discussed the role of the voluntary sector in contributing to the social and cultural inclusion of older dependent people. We gave examples of some good locally based models of care provision that provide opportunities for sociability and meaningful involvement.

Chapter Four
Leisure and Social Activities

Introduction

This chapter examines social participation through leisure activities, involvement in social clubs and informal sociability. The focus is on the ways in which participation in leisure promotes connectedness. We explore the types of interests and social activities that are most associated with intrinsic satisfaction and that express and encourage friendships. Whether leisure is an important source of meaning depends on the intersubjective experience of those involved in the activity. We wished to know, therefore, whether and in what ways the social relations that people experienced through leisure were a source of meaning. To what extent was a sense of involvement with others intrinsic to the enjoyment of leisure? How did particular types of leisure activities influence how older people experienced the world? How did those who appeared to have few leisure interests or whose opportunity to participate in activities was limited by health problems experience meaningful interactions?

At the outset, we were interested in finding out to what extent the leisure interests and social participation of older people were locality based or geographically dispersed. General affluence and easier transport makes it possible to pursue hobbies and interests away from one's locality. Dispersed family networks mean that intimate relationships are often maintained at a distance. Entertainment can be accessed in one's home through television and other forms of mass media. Religious activity may be a private affair and can be engaged in through a variety of institutional means and not necessarily in one's locality. Friendships may also be maintained at a distance through arranged meetings, visits and telephone contact. More privatised and intimate relationships are said to be a feature of post-modern society. Therefore, considering all these factors, it was quite possible that older people would live contented lives without many significant relationships in their immediate localities and possibly without any involvement in formal or informal groupings outside their immediate family. However, our investigation suggested that social interactions in one's neighbourhood are of immense significance in the lives of older people. The overall argument in the chapter is that through communal leisure activities people create social worlds that make life meaningful and worthwhile.

The chapter begins by considering the nature of leisure and particularly perspectives on leisure in retirement. We present a brief profile of the respondents in terms of their retirement status, their satisfaction with retirement and their broad sense of whether they have enough to occupy and interest them. We map out the most significant social groupings in the lives of older people in Rathmore and in Rathbeg. These include sports groups, social clubs, Active Retirement groups (ARAs), special interest groups, voluntary and community groups and church groups. Patterns of participation are described and differences in terms of gender, age, social class and location are highlighted. The most common and the most individual leisure activities of the respondents are described and classified, for example common home-based activities, outdoor physical activities, special interest activities and continuing education. The major interests of our respondents are described and discussed: reading, watching sport on television, participating in sport, gardening, art and craft work, family related work, current affairs, music, voluntary commitments, television and travel. The social and personal significance of leisure activities are examined, drawing on theories and concepts related to lifespan development and the Third Age.

Leisure and social participation

Leisure can be seen as a site of enjoyable activity that acts as a contrast to those compulsory and regular activities that must be undertaken for survival and maintenance – activities thought of as work. However, the distinction between work and leisure is not at all clear cut and many activities can be defined as either work or leisure or both insofar as they have the potential to provide the means through which people can express themselves and gain enjoyment. Seeing leisure as doing something that one chooses to do allows a wide range of activity to be included such as voluntary work, quiet home-based activity, educational and religious activity, hobbies and special interests.

Aristotle viewed leisure activity as providing opportunity for self-development when the necessities of life had been taken care of. Indeed he appeared to argue that what people do in their leisure defines the type of person they are. However, not all forms of leisure have the dimension of growth or self-development. Some leisure activities, for example doing art, may give us a more creative perspective on life as well as contributing to self-development. Other activities such as watching television tend to be more passive and less demanding.

Leisure in retirement

Writers such as Moody have pointed to competing views of leisure associated with retirement. In a narrow sense leisure can be seen as time away from work, as discretionary time which becomes more available in retirement. However, Moody suggests that leisure is more than 'leftover time', rather it is 'a multidimensional quality of life different from paid employment, household maintenance, and other instrumental activities' (Moody, 1998: 438). This more positive view of leisure which has also been proposed by other writers such as Kaplan (1979) sees leisure as offering opportunities for recreation, relaxation, personal development and service to others. Indeed, Moody argues that in many ways the alternative of work versus retirement is a false choice. He suggests that the real question may involve how to enable older people to live lives of greater productivity, whether in employment or in retirement (Moody, 1998: 326). The question of what is productive ageing or successful ageing is, in fact, an underlying theme throughout the book.

In retirement many older people have more time for leisure and a less structured work/leisure division in their daily routines. However, paid work is a major source of social status, self-esteem and social contacts in our society. Moody suggests that what is problematic about retirement is not that work is always better, but that the abundance of free time in late life is not adequately structured for any larger social purpose or meaning (Moody, 1998: 321). A more pessimistic view has seen the exit from paid labour as segregating older people from others in society and contributing to perceptions of older people as socially redundant and a burden (Phillipson, 1982; Featherstone and Hepworth, 1989).

However, while in retirement many older people have more time for leisure, activities involving leisure or voluntary work suggest many different orientations. For example theories of the Third Age tend to emphasise personal achievement and fulfilment through creative and satisfying hobbies and interests. Laslett views the Third Age as a time to do the things that we have always wanted to do without the trammels of the Second Age (Laslett, 1996: 154). However, this assumes that individuals will have developed interests and hobbies at an earlier stage in life and that they pursue them enthusiastically in retirement and/or that they are willing to try out new activities. Furthermore, as political economy perspectives remind us, such involvement in creative leisure takes place within contexts strongly affected by socio-economic structures and the circumstances of exit from the workplace.

Public discourses on retirement and leisure tend to emphasise individualistic goals such as having a good time, self-fulfilment and individual choice. These seem to be associated with acquisitive individualism and consumerism to the extent that they contain little reference to altruistic values or the contribution that older people can make to society. Yet, we found that many of the social activities of older people are directed towards helping others. This suggests yet another orientation that emphasises civic virtue and the satisfaction that can be derived from altruistic activities.

We will therefore examine leisure and social activities to see whether and in what ways they provide mutuality of experience and provide opportunities for the older adults in our study to continue to engage in reciprocal intergenerational and communal relationships.

Work, retirement and leisure patterns of respondents

The overwhelming majority of the respondents were no longer involved in paid work. Almost 73 per cent of the Rathmore respondents described themselves as retired, 20 per cent, all of whom were female, described themselves as engaged on home duties, and six said they were unable to work because of illness. Five respondents described themselves as employed while ten said they had done some paid work in the previous year. When asked whether or not they had been happy to retire, two-thirds said they had been happy to retire, while the remainder said they were unhappy or had mixed feelings.

It was not our intention to unravel the complexities of the relationship between expectations of retirement and experiences of retirement. Where retirement was involuntary it was not surprising that while respondents may have adjusted to it, they did not see it in a positive way. Feelings of resentment were common among people who were forced to retire because of redundancy or ill health: 'I was forced to retire at 68 due to an accident … I felt cheated, I loved my work.' 'I was forced to retire at 62, I felt I could have gone on more but the factory closed.'

Some respondents including women who worked at home since they got married said that retirement had not changed anything for them – they were still doing the same things. However, dissatisfaction with retirement was also found among respondents who had retired at the normal retirement age. Some commented that they missed their workmates, or they had got sick when they retired or they had been active for a number of years but were now less active: 'I'd rather still be working. I feel lonely without the people I used to work with.'

Several of the respondents said that they saw retirement as a new phase in life or a different way of life. Many of the respondents who were positive about retirement lived what could be described as 'full lives' exemplified by a variety of social activities and commitments including leisure interests, community and voluntary activities and involvement with children and grandchildren. Some spoke about the freedom to do things and having reduced responsibilities:

> Having time to do a lot of things I never had time to do. I have more freedom. I'm enjoying life now, relaxing, taking it easy. Retirement has made me twenty years younger … no worries or responsibilities.

Respondents who were members of clubs or groups, those who had a major interest and those who did voluntary work were more likely to be positive about retirement. These findings agree broadly with those of Whelan and Whelan who found for example that 41 per cent of respondents who had a hobby were enjoying retirement very much compared with 13 per cent of those without a hobby. In relation to voluntary work, 78 per cent of those who were involved in voluntary work were enjoying retirement compared with 59 per cent of those who were not engaged in such work (Whelan and Whelan, 1988: 39).

Respondents were asked to select a statement that best described their sense of whether they had enough activities to fill their day.[1] None of the respondents selected the statement: 'I have nothing to do.' Six respondents chose the statement: 'I have little to do, I have a lot of time on my hands.' A further 11.5 per cent chose the statement: 'I have plenty of free time', implying that they would like more to do. In contrast to these respondents who did not have enough to occupy them, over one quarter of respondents classed themselves as 'not having enough time in the day and being very busy'. The most common response selected by 57 per cent of the respondents: 'I have a full day but not too much to do' suggests a sense of balance, of having meaningful activities but not being pressurised. The six respondents who said they had little to do all lived alone, were predominantly from the manual social classes, and were either single or widowed. Respond-

1 This question was taken from the Eurobarometer survey 1993.

ents who said they were very busy were often caring for a dependant in their own household and/or involved in helping someone outside the household and/or doing voluntary work. However, we will explore patterns of engagement more fully later.

Several of the women said that they kept themselves occupied with housework and cooking while a few men said they did all their own house maintenance. A few men spoke about how they enjoyed the freedom that retirement had given them to take off on public transport to the sea or 'down the country'. However, even respondents who had hobbies and interests could experience boredom at times. One female respondent reflected on this: 'With old people you don't do things in hours ... when the weather is bad I mope a lot.' We will now go on to discuss the group involvement and leisure activities of older people in Rathmore and Rathbeg.

Mapping social groupings in Rathmore and Rathbeg

Because of the complexity, diversity and interrelatedness of social groupings we approached the task of mapping and identifying the significant social groupings at a number of levels. In Chapter One we described the methods used to obtain knowledge of social groupings and settings and patterns of social participation. We have divided our findings on social participation into those whose primary purpose is leisure and enjoyable social activities and activities thought to be more connected with engagement in civil society through voluntary activity, religious practice and involvement with community groups. This chapter examines social participation through leisure activities, social clubs and informal sociability while Chapter Six will examine voluntary work, community development activities, church based activities and educational activities. While we separated out groups in terms of whether they were primarily concerned with leisure, voluntary work, community development, religious activity or educational activity, in practice these collective undertakings are often interlinked.

Configuration of groups

On the basis of our ethnographic, survey and interviewing strategies we identified the following groups that were active in Rathmore and Rathbeg and that seemed of particular relevance to older people's lives:

Groups in Rathmore

- Sports clubs: GAA clubs, soccer clubs, golf and other sports clubs.
- Social clubs: Senior Citizens' clubs, Ladies' clubs and Recreation clubs.
- Active Retirement groups.
- Church related groups: including Lay Ministries, Parish visitation groups, church maintenance work and church choir.
- Community development groups: including Community Development Projects funded by state agencies, Credit Unions, Residents' Associations and Neighbourhood Watch.
- Special interest groups: including art groups, historical society, reading club, music and dancing groups.
- Charitable groups /Voluntary work: including the St Vincent de Paul Society,[2] Senior Citizens'/Old Folk's committees (meals on wheels provision and volunteering in day centres and residential nursing homes).
- Day centres for older people.

2 The St Vincent de Paul Society is an International Voluntary Christian organisation whose ethos is based on the gospels and Catholic social teaching. There are 9,000 members in Ireland in 1000 conferences organised on a parish basis across all 32 counties. The Society describes its work as having three components: support and friendship, promoting self-sufficiency and working for social justice (Society of St Vincent de Paul official website).

Groups in Rathbeg

In Rathbeg and its environs we found a similar configuration of groups with some differences:

- Sports clubs: soccer club and golf club.
- Social clubs: Irish Countrywoman's Association,[3] women's group.
- Church related groups: similar groups to Rathmore.
- Community development groups: Local development groups, Community Development Projects funded by state agencies, Community Alert, Credit Union.
- Special interest groups: examples include art group and historical society.
- Charitable groups /Voluntary work: including the St Vincent de Paul Society and Care of the Aged committee.
- Day centres for older people.

While there was a similar configuration of groups in both areas, there were important differences. There is no Active Retirement group in the immediate environs of Rathbeg. However, there are Active Age groups in other parts of the County. There is a soccer club but no GAA club in Rathbeg. Not surprisingly there was a far greater number of groups in Rathmore and more community and parish centres where communal activities could be held. One parish directory in Rathmore listed 44 ministries within the Parish.[4] A local newspaper[5] listed 23 community groups in Rathbeg. However, not all the groups are equally

3 Founded in 1910 the Irish Countrywoman's Association has branches, called guilds, throughout Ireland. There are 47 guilds in Dublin City and County and 29 in County Donegal. Traditionally the ICA was associated with craft activity, however interests have broadened in recent years (ICA official website). In 2004 the Rathbeg branch of the ICA disbanded due to dwindling numbers. Interestingly, around this time a new local women's organisation was established. According to a Directory of Local Organisations the group's motto is: 'Friendship, fun and support'.

4 A Rathmore Parish directory, 2003. A parish centre was opened there in 1997.

5 Newspaper dated October 04 sold in Rathbeg and environs.

active in Rathbeg. For example three or four of the groups listed organise a relatively small annual event and would involve only a handful of people. Other groups, for example charitable groups or community development groups would meet regularly. Some expend considerable time and effort in for example developing local amenities or organising a local festival and each group would involve between five and ten active members. We also observed important differences in the voluntary groups in both areas as discussed in the previous chapter.

Membership of clubs and groups in Rathmore

Being involved in a club or a formal social group is one indicator of connectedness and may provide an important source of meaning for older adults. From a list of ten different types of groups and organisations (Q. 6b in questionnaire) respondents were asked about their involvement if any, both present and past, with each type of group. We included different types of formal groups – sports, charitable and fund raising, church based, community development, Ladies' clubs, Active Retirement, adult education, special interest groups and political parties. Approximately half of the respondents in Rathmore were currently involved in at least one group or organisation while almost 60 per cent were either presently or previously involved. Table 4.1 below shows the proportion of respondents currently involved with each group on a regular basis. We recognised that 'regular involvement' can range from helping with a specific task to taking a leadership role, but we are not concerned with such distinctions at present. The most prominent groups were church based (15.2 per cent of the sample) and sports based (14.5 per cent). These were followed by community development groups (9.1 per cent), charitable groups (6.7 per cent) and Active Retirement groups (6.7 per cent). The numbers who actually participated in ARA activities was higher than this however. We noted that, in response to Q. 21 on participation in leisure activities, 9.1 per cent said they participated in social activities through an ARA. Other groups with whom the respondents had regular involvement included fund raising groups (4.8 per cent), adult education groups (4.2 per cent),

Ladies' club (2.4 per cent) and music/drama groups (2.4 per cent). Only one respondent (0.6 per cent) was an active member of a political party.

Table 4.1: Clubs and groups that respondents were involved with on a regular basis

Club/group	% of respondents
Church	15.2
Sports	14.5
Community development	9.1
Charitable	6.7
Active retirement	6.7
Fund raising	4.8
Adult education	4.2
Ladies' club	2.4
Music/drama group	2.4
Political party	0.6
n	165

Approximately half of the respondents in Rathmore were currently involved in at least one group or organisation – 23 per cent were involved in one group, 19.4 per cent were involved in two while 7.2 per cent were involved in three or more groups. Men were more likely to be members of groups than women but the difference was not statistically significant and was largely accounted for by the fact that men are much more likely to be involved in sports organisations than women. Men were three times more likely to be involved in a sports club than women. The older age group in the sample were less likely than the younger age group to be involved in a group. Of those over 75 years, 63.3 per cent were not members of any group while just 39.3 per cent of those under 75 years were not members. Of those involved in two groups or more 31 per cent were 75 years or older while 69 per cent were under 75 years. Those under 75 were twice as likely to be involved in 2 groups or more as those over 75.

Some people gave reasons why they were not involved in groups. The main reasons given referred to constraints due to health and mobility difficulties and lack of interest or desire:

> 'I used to do a lot more, I'm not really able now.'
> 'I'm not involved in anything since I got my hip done. I'm a background sort of person.'
> 'I never had an interest in those.'
> 'I keep to myself'.
> 'I was too busy raising a family – I'm a home bird.'

At the same time a few respondents expressed an interest in getting involved in something 'to get out of the house'.

Rathbeg

Turning to Rathbeg we found that older people were involved in similar groups including St Vincent de Paul, fundraising for church and other charities, local enterprise and development groups, ICA, political party and special interest groups such as an Historical society and an art group. A different type of group membership that we came across was the Royal British Legion; one of our respondents was recently awarded a CBE medal for his contribution to the welfare of war veterans of the British army. We noted that respondents who had leadership roles or who served on committees in clubs or in community development projects had generally been involved in such activities over a long period. However, some had become involved on retirement and had put their professional and work skills to use in a particular community project. People who preferred a less public role could find opportunities to be involved as ordinary social members of clubs or could be found helping at specific community events or in their local church. While only a minority of older people in Rathbeg are active members of clubs, many attend social events such as fundraising coffee mornings, church events, organised outings and other community events.

Findings from other studies on participation in club/group activities

The literature suggests that broadly speaking as people get older they usually engage in the same activities they did earlier in life, but with advancing age there tends to be an overall decline in participation rate (Moody, 1998: 439). A recent national survey on income, living standards and quality of life (Living in Ireland survey 1997) found that one third (33.5 per cent) of those aged over 65 living in urban areas were members of a club or organisation while one fifth (19.5 per cent) of older people living in rural areas were members (Layte *et al.*, 1999: 124). However, this survey did not include church based groups. Given that their specific definition of clubs and groups did not include church groups and that about 15 per cent of the present sample were members of church based groups the findings in relation to Rathmore seem fairly consistent with those in the Living in Ireland survey.

In relation to gender differences the 1997 survey found that men were significantly more likely than women to be members of clubs and organisations in all age groups and irrespective of location, apart from among young elderly people (those aged 65–69) in rural locations where women had higher participation in clubs and organisations than men. The authors pointed out that while there was some evidence that the gender gap closes with age, membership of clubs and organisations declines with age. The authors commented that while it might appear understandable that older people are less likely to play sports that they may have played when they were younger, they found it interesting that older people did not appear to replace membership of clubs with new affiliations (Layte *et al.*, 1999: 124). Perhaps this is not so surprising since we found that membership of the significant social groups in our study areas appeared to represent not just an instrumental activity but involved commitments and relationships that were meaningful and took time to evolve; replacements might not be readily available. One does not casually pick and choose social groups as one would choose a consumer purchase. On the other hand if given opportunities to experience new activities people can find to their surprise that they enjoy them.

Studies have shown that while some activities decline with age, others remain the same. A Canadian study found that taking up new activities diminishes as people get older; cinema going, travel and indoor fitness also decline significantly. However, other activities such as gardening show only modest decline while church participation and community activities tend to be maintained (Iso, Jackson and Dunn, 1994). While people discontinue involvement with social clubs for different reasons and at different phases of their lives, and while clubs themselves wax and wane it seemed that there were continuities in our respondents' involvement in certain groups, in particular church groups and community/voluntary groups.

Older people's social groups and clubs

Active Retirement groups and Senior Citizens' clubs provide opportunities specifically targeted at older people to socialise in their mid to later years. Other clubs and social settings that are important meeting places for older adults are sports clubs, Ladies' clubs, special interest groups and pubs. Active Retirement groups and Senior Citizens' clubs present somewhat contrasting models of participation.

Active retirement groups

The advent of Active Retirement groups is relatively recent compared with Senior Citizens' clubs.[6] While ARAs are more likely to be formed in middle class rather than working class areas, there is some evidence that ARAs may be taking over from Senior Citizens' groups in urban

---

6 The Active Retirement Movement was founded in Dun Laoghaire in Dublin in 1978 and has grown since then to a National membership of 15,000 comprising over160 Active Retirement Associations (2005). Each Association is autonomous in keeping with the self-help ethos of the Movement and the organisation is based on a network of associations affiliated to the Federation of Active Retirement Associations (FARA). Associations are either locality based or attached to a work place. Membership is open to people aged 55 and over who have retired from their main occupation or whose families are largely reared.

areas.[7] In our survey in Rathmore we found that 15 respondents (9.1 per cent) took part in ARA organised activities while six (3.6 per cent) persons participated in Senior Citizens' clubs and four (2.4 per cent) in Ladies' clubs. The aims of ARAs are very social in nature and include promoting lifelong learning, emphasis on the contribution of older people and seeking to combat social isolation.

We can look more closely at one of the four Active Retirement Associations in Rathmore which started in 1989 and which has a membership of 120 persons. The Association is self-financing, the organisation of activities is done on a voluntary basis and they receive a small grant from Dublin City Council for bowling equipment. They advertise their activities in the Parish newsletter, by word of mouth and at meetings and other events. Activities include movement to music, sequence dancing, bowls, art, bingo and cultural outings. The Secretary of the ARA said:

> The Active Retirement Association is the most wonderful thing ever. It doesn't matter what age you are … we're not conscious of each other's age. You can be young at 80 or old at 50 in your mind. Take men: they work, retire, may not have that many hobbies – it can be a 'let down' to retire (Interview with Secretary of ARA 17/09/03).

Apart from a sense of personal achievement that accompanies participation in some project or physical or creative activity, the potential of an Active Retirement Association to affect people's lives in many other ways was evident in the interview with the Secretary:

> Some people just like to feel part of the group, just to sit and listen and not feel lonely. They would go on trips. There are people who are not outgoing – you have to respect their privacy (Interview with Secretary of ARA 17/09/03).

This emphasises that an Active Retirement group can be flexible and responsive to different interests of older people in a locality, even those who would not be described as active or outgoing. A strong sense of social concern and kindness to other members was evident

---

7 Membership of Ladies' clubs which are more commonly found in urban centres and the Irish Countrywoman's Association which has branches throughout the country has also declined in recent years.

when the Secretary spoke about the sensitivity and support displayed towards members experiencing some diminution of cognitive capacity:

> People in the Active Retirement Association are supportive and protective of those developing Alzheimer's. They look after them on outings, holidays, to the theatre. You never hear them complaining about having to do this. I think people know what you're going through (Interview with Secretary of ARA 17/09/03).

An Active Retirement Association can be a conduit for older people to take on a productive role in their communities. An interesting initiative organised by a local National School in Rathmore involved volunteers from the Active Retirement Association becoming involved in a Paired Reading project to help the children with reading. The school used the opportunity of the Paired Reading initiative to develop good intergenerational understanding between the pupils and the older volunteers through, for example 'question and answer' sessions where the pupils asked the volunteers what they did as children and the school children hosting a meal for the older volunteers. Intergenerational school-based programmes have been developed in the United States, Europe and Australia and research evaluation has suggested that they have much to offer (see for example Bales *et al.*, 2000, Lubarsky, 1997, Feldman *et al.*, 2003). The Secretary of the Active Retirement association said: 'It teaches the children to have respect for older people' (Interview with Secretary of ARA 17/09/03).

On a national level the work of one ARA in Summerhill, County Meath has led to the development of a national service, the Senior Help Line, to provide a telephone listening service for older people who are lonely and isolated. The Summerhill ARA set up the Senior Help line with the support of the Health Boards, the Samaritans who provide training and a network of 300 older volunteer. The service has expanded from one centre at Summerhill to a network of centres providing a service seven days a week in response to a steadily increasing number of calls (O'Shea, 2003: 103).

The work of the Rathmore and other ARAs demonstrated the capacity of such groups to encourage and promote meaningful and enjoyable activities in a flexible way and with a strong ethos of social concern. What is particularly interesting is that the Association seemed to be able to respond and cater for those who wanted active participation and those who mainly wanted company. Active Retirement Associations are more common in urban areas than in rural areas where members can meet regularly in a convenient location usually within a Parish area. However, in rural areas including County Donegal Active Age groups have been forming in the past ten years.[8]

Senior Citizens' clubs

A more traditional type of social club for older people is the model of the Senior Citizens' club sometimes referred to as 'The Old Folk's' committee or club and which can be contrasted with the Active Retirement Movement with its self-help and active participation ethos. In Rathmore a number of Senior Citizens' clubs have operated since the late 1960s with a predominantly female membership. They have experienced mixed fortunes. There is one very active group in Rathmore, a second group has experienced a decline in its activities and a drop-off in the number of participants, while a third Senior Citizens' club has recently disbanded.

The active Senior Citizens' club which is multi-denominational and open to anyone over 60 years organises regular social activities that provide fun and entertainment for many older people in the locality. They meet one or two evenings every month for social activities such as music evenings, bingo, trips to a cultural event or activities to mark a seasonal Festival. In addition there are annual day outings, trips away and a Christmas party. The events organised by the Senior Citizens' club are undoubtedly popular given the high levels of participation, with numbers from 100 to 150 participating in outings and attending functions. The Secretary who is the main driving force in the club is herself in her 80s and has been actively involved in the com-

8 In rural areas Active Age groups seem to be the preferred form of organisation as ARAs can be seen as too professionally based. It would appear, however, that there is no major difference in the philosophy of Active Age groups and ARAs. (Information from Health Promotion Officer in Donegal).

mittee for over thirty years. She organises a very full programme of events each year and her energy and commitment is admired and respected. She has said of her work: 'I am the committee … it's more or less my life.'[9] While there is a small, dedicated committee who help, it has been suggested that the committee may be too closely identified with the Secretary herself and that the committee is too 'closed'. On the other hand people greatly admire the tenacity and energy of the main organiser who has been described as 'tough' but also 'great for getting things for the older people'[10] and 'a bit autocratic but a great organiser.'[11]

The decline of traditional groups providing entertainment for older people was observed in the second Senior Citizens' club in Rathmore.[12] The organisers, all in the older age group themselves, talked somewhat nostalgically about activities that they used to run with far more participants – activities such as drama, concerts, home visitation and door-to-door fund raising. One of the organisers has been involved in the club/day centre for 27 years and the last volunteer joined ten years ago. Most of the club participants are residents of the sheltered housing nearby while the committee of about 5 or 6 women live in the general surrounding area. There was a realisation that the club was no longer as popular. This was discussed in terms of changes such as many members having died, residents of sheltered housing being better off now and more older people having cars. One committee member remarked: 'I suppose this will finish too but it would be a shame'. Conditions of life have changed since these women became involved as volunteers in the Senior Citizens' club. Becoming involved was an important social outlet for women whose opportunities to participate in activities outside the domestic role were limited. One woman who had only become involved ten years ago and who got great personal satisfaction from the work said that her only regret was

---

9 Interview with the Secretary 12/3/01.

10 Conversation with an informant 16/9/03.

11 Interview with an informant 12/3/01.

12 Social activities take place on two afternoons each week in a Social Services Centre in Rathmore, bingo on one day and music, dancing and exercises on the other.

that she was not asked to get involved at an earlier stage in her married life.

This more traditional model of providing entertainment, organising social outings and fund raising in a locality for 'The Old Folk' can be contrasted with the Active Retirement Movement. Furthermore, the strength of traditional clubs, often residing in the commitment and personality of people who are the driving force behind them, may also militate against the long-term continuity of such groups. The structure of ARAs is more democratic, provides more flexibility and choice in relation to activities, and ARAs have an identity not closely associated with any one person. Another contrast is that the Senior Citizens' clubs are largely female dominated while ARAs tend to attract men and women.

From a social policy perspective it has been observed that even well established day centres have experienced a decline in membership in recent years with fewer new members replacing those who have moved in to residential care or who have become incapacitated or who have died. One recent study suggested that changing expectations of the 'younger old' make them less inclined to make use of services as currently arranged. The study suggests factors such as improved health, higher income and the increasing role played by grandparents in child care as impacting on participation (O'Sullivan, 2003: 9).

Leisure interests and hobbies in Rathmore

From a list of 31 different hobbies and leisure interests respondents were asked whether or not and how often they engaged in each activity (Qs.21a and 21b). Table 4.2 reveals the most common and most individual leisure activities of older people in Rathmore.

Table 4.2: Hobbies and leisure interests of respondents in Rathmore

Hobbies and interests	% of respondents
Television	96.4
Newspapers	92.7
Listen to music	81.8
Radio	81.2
Sport on television	73.9
Gardening	69.7
Reading	69.1
Walking/physical exercise	67.9
Pub	36.4
Library	35.8
Crosswords	30.9
Outdoors sport	23.6
DIY	20.6
Art and craft	19.4
Concerts	19.4
Playing cards	17.6
Use a computer	12.7
Cinema	12.1
Bingo	10.3
Active retirement association	9.1
Theatre	8.5
Indoor sport	6.7
Creative writing	6.1
Adult education	5.5
Playing music	4.2
Swimming	4.2
Senior citizens' club	3.6
Dance	3.6
Choir/music group	3.0
Day centre	3.0
Ladies' club	2.4
Scrabble	2.4
n	165

In the discussion below we have grouped the activities in order of frequency of participation. However, we have also tried to combine different categories of leisure interests together such as sport, art/craft work and attending Arts entertainment.

- Activities engaged in by at least two thirds of respondents were watching television (96.4 per cent), reading newspapers (92.7 per cent), listening to music (81.8 per cent), listening to radio (81.2 per cent), watching sport on television (73.9 per cent), gardening (69.7 per cent), reading books (69.1 per cent) and walking (67.9 per cent).
- The next most popular activities which were considerably less common than the above activities involving participation rates of about one third of the respondents were: going to the pub (36.4 per cent), using the library (35.8 per cent) and doing crosswords (30.9 per cent).
- Approximately one quarter of the respondents (23.6 per cent) played an outdoor sport, most typically golf, while only 6.7 per cent played an indoor sport such as snooker. Swimming was another physical leisure activity which seven respondents participated in.
- Activities related to art/craft work or DIY were engaged in by approximately one fifth of the respondents (19.4 per cent and 20.6 per cent respectively).
- In relation to interest in music and the Arts about one fifth said they went to concerts (19.4 per cent), about one eighth of respondents (12.1 per cent) went to the cinema while a smaller proportion went to the theatre (8.5 per cent).
- A smaller proportion of respondents were regular participants in music, singing and dance groups – seven people played a musical instrument, six people took part in dancing while five people sang in a musical group. Another creative activity, creative writing, was an interest for ten people. No one in the sample was involved in a drama group.
- Playing cards was a pastime in which 17.6 per cent of the respondents engaged most usually in their own homes or the homes of friends or relatives. Bingo which is often popularly

associated with older people was in fact a pastime for only 10.3 per cent while scrabble was played by four respondents.

- Approximately 16 per cent said they participated in social activities centred on membership of a club or social group for older people: fifteen persons or 9.1 per cent were involved in an Active Retirement Group, six were involved in a Senior Citizens' Club while five were regulars at a Day Centre. While Ladies' clubs cater for women of different age many members are in the older age group. Four women in our sample were members of a Ladies' Club.
- In relation to adult educational activity nine people were involved in formal education classes. Using a computer was an activity engaged in by 12.7 per cent of the sample – more, therefore, than played bingo.

Some activities such as playing an outdoor sport were predominantly a male interest while activities such as art/craft work were predominantly a female interest. Bingo was overwhelmingly a female interest; 16 female respondents and only one male respondent said they played bingo. On the other hand participation in an Active Retirement group was not differentiated by gender .

Major interests

The respondents were asked if there were other leisure activities that they participated in that were not listed in Question 21a and the answers to this question revealed an even wider variety of interests, such as dominoes, researching family history, hill walking, jigsaws, bird watching and photography. However, if these 'other interests' were particularly important in the lives of respondents, they had an opportunity to express this in Question 21c where the respondents were asked whether they had one or two major interests. We found that in many cases these more idiosyncratic interests were also major interests and frequently involved social interaction. One man who played competitive draughts and competed with other clubs once every week described this interest as 'serious fun'. Another male respondent referring to his involvement in a Credit Union described it as a major interest. A female respondent who had a major interest in bird watching but who was unable to pursue it as actively as she used to spoke about having continuing contact with friends and their children whom she got to know through this interest.

Major interests: Rathmore

Just over three quarters said they had a major interest while just under a quarter had not. A small number of those who had no major interest said that they 'were interested in lots of things' or that they 'looked at everything but never got carried away with anything' or that they liked a 'balance of interests'. The most common major interests were:

- reading,
- watching sport on television,
- participating in sport, and
- gardening.

The other major interests in rank order were:

- current affairs,
- family related work,
- art and craft/DIY activities,
- music,
- voluntary commitments,
- travel, and
- television.

Tables 4.3 and 4.4 below show the respondents' first and second major interests. We found that reading was a first major interest for 10.3 per cent of the sample and a second major interest for 6.1 per cent. Sport was a first major interest for 21.2 per cent – this included 10.3 per cent who said that participating in sport was their major interest and 10.9 per cent who said that following sport was their major interest. The most common sport played by respondents was golf, 8 per cent of the sample having selected it as their first major interest. The main sport that people followed was soccer. Other sports followed included rugby, Gaelic football, hurling, golf and horse racing. Gardening was selected by 7.9 per cent as their major interest while 3.6 per cent chose it as their second major interest. The other major interests selected in first place were current affairs (4.2 per cent), family related work including cooking and caring (3.6 per cent), Art and craft/DIY activities (3 per cent), music (3 per cent) and voluntary commitments (3 per cent). Travel and television were chosen in first place by 2.4 per cent and 1.8 per cent respectively.

Table 4.3: First major interest of respondents

First major interest	% of respondents
Reading	10.3
Sport on television	10.9
Participating in sport	10.3
Gardening	7.9
Current affairs	4.2
Family/cooking/caring	3.6
DIY/art/craft	3.0
Music	3.0
Voluntary commitments	3.0
Travel	2.4
Television	1.8
Other	16.6
No major interest	23.0
n	165

Table 4.4: Second major interest of respondents

Second major interest	% of respondents
Reading	6.1
Sport on television	2.4
Participating in sport	1.8
Gardening	3.6
Current affairs	2.4
Family/cooking/caring	3.0
DIY/art/craft	1.8
Music	3.0
Voluntary commitments	1.2
Travel	0.6
Television	1.8
Other	5.6
No second major interest	66.7
n	165

When we crosstabulated first and second major interests with gender and age we found that participating in sport was overwhelmingly selected by males as a major interest. Watching sport on television was also largely a major interest of males. In contrast, reading was overwhelmingly chosen by females as their major interest. However, more males chose reading than females as a second major interest. Gardening was chosen by more females than males as both first and second major interest. Females were also more likely to choose art and craft type activities and home making/family interests and they were somewhat more likely to choose current affairs. There were no striking gender differences in choosing voluntary commitments or music as a major interest.

Participating in sport was a major interest that was more predominant among those in the younger (under 75) age group. However, while indoor sport was only chosen as a major interest by four respondents, all were in the older (over 75) age group. There were no striking age differences in relation to choice of other activities as a major interest. While reading was chosen as a first major interest by almost twice as many in the older age group, it was chosen as a second major interest predominantly by the younger age group. In relation to watching television it was people in the older age group who chose it as a first major interest while people in the younger age group chose it as a second major interest. There was a tendency for the older age group to choose current affairs more than the younger age group.

Major interests: Rathbeg

Turning now to our more rural community the smaller survey of eight older people in Rathbeg revealed a similar pattern of leisure pursuits including reading, sport, gardening, walking, art, sewing, pub, computer, cards, theatre, bingo and DIY. While most of these interests could be practised at home or locally, some such as going to the theatre or bingo or doing a computer course would involve a car journey. The major interests of the Rathbeg respondents revealed a similar pattern of common and special interests as we found in Rathmore:

- Soccer – three persons mentioned this and one added sport in general (two of the respondents had played soccer when they were younger)
- Voluntary work
- Meeting people/social awareness
- Gardening and house
- Reading and crosswords
- Heritage project

One difference between Rathmore and Rathbeg is that in Rathbeg people have to have private transport to attend cultural events or to enrol in classes. We noted that respondents who could drive and generally those who had a higher level of formal education tended to avail of wider cultural facilities in the nearby larger towns where theatre, libraries, exhibitions and other cultural activities took place. On the other hand the establishment of a local art group and a local

Historical society have opened up more possibilities for people to develop their own creativity and to deepen their knowledge and appreciation of their locality and their heritage.

Card playing and bingo are social activities enjoyed by mainly middle aged and older people in Rathbeg. Informal card groups of four people (usually men) meet regularly to play card games such as 'solo' in each other's houses during the winter. Cards are also played in venues such as pubs usually as a fund raising venture, involving games such as 'twenty five' or 'whist'. It appears, however, that card playing has dropped off from the level of card playing in the past. We found in our survey in Rathmore that playing cards was a pastime that 17.6 per cent of our respondents said they did. However, card playing did not emerge as a major interest for any respondent and it often involved a fun game with grandchildren. While men played cards in a more serious way in venues such as sports clubs, women tend to play with friends in each other's homes.

Common home based leisure activities

Reading, television, radio and crosswords which are common pastimes for older people are clearly part of the texture of everyday life for individuals of all ages and take place largely in the private life of households and families. Such activities serve a number of purposes; they keep people in touch with the world, they provide entertainment and they keep people mentally active. Television, radio and newspapers are accessed on a daily basis by the older people in this study. Our respondents are keen newspaper readers and the vast majority buy either a morning or an evening newspaper. Newspapers provide coverage of two areas of particular interest to our respondents – sport and current affairs. A few said that they also read the regional newspaper of their county of origin for example 'the Kerryman'.

An important purpose of leisure among older people is to keep their minds active and some respondents spoke about how they purposely chose particular activities in order to keep their minds alert, for example crosswords, scrabble or bridge. One lady who celebrated her one hundred years birthday during 2004 does crosswords regularly. She said she also does mind games to keep herself alert such as saying all the surnames beginning with 'Mac' that she knows. This lady also takes part in quizzes in the local day centre and generally wins first prize. Another lady said she sometimes played scrabble on her own and said that she hoped such activities would prevent dementia.

Television and radio

Home based activities can be especially important for people whose mobility is restricted, who live alone or who don't have opportunities or don't seek opportunities to engage in more collective or more creative leisure activities. While home based leisure activities may involve deep interests, for example in literature or current affairs, much of television viewing involves more passive entertainment. Many respondents spend several hours watching television every day primarily in the evenings. However, many of our respondents spoke about 'selective viewing' of particular types of programmes. Some respondents expressed a preference for the 'soaps', others for news and current affairs, while others enjoyed documentaries, films such as old Westerns, comedies or cookery programmes.

Several respondents who lived alone referred to television or radio as 'company in the house'. One respondent who liked to have television on during the afternoon said that television 'brought life' and 'livens up people.' While television viewing is generally seen as an individual activity there can be a social aspect to it, when people view programmes together. This is particularly evident in relation to big sporting events such as major soccer, rugby and Gaelic matches when family members sit down together and lively commentary and friendly argument ensue. We came across companionable viewing of sporting events by family members including husbands and wives and son and father-in-law. According to one study television is a crucial medium for helping older people keep in touch with the world (Gauntlett and Hill, 1999).

While radio listening was not as common as television viewing many respondents spoke about the substantial amounts of time they spent listening to the radio and how much they enjoyed it. Several respondents said they listened to the radio all day, particularly RTE Radio 1 where they followed all the current affairs and light entertainment programmes. Others liked to listen in the morning or while doing their household chores. Some spoke about how they liked to listen to music, for example on Lyric FM. An indication of how radio could also serve as 'company' was given by several respondents who said how much they valued the radio at night: 'I listen every night and if I waken up it's handy to have it near you.' In Rathbeg local radio is also listened to for local news, local events and obituaries.[13]

Reading

Reading emerged as the most common major interest among the participants in the Rathmore survey, and over one third of respondents said they used the library. Some mentioned their particular reading interests such as history, Cowboy and Western, bridge books and poetry. A number of people said that they like to read religious books and magazines such as the Messenger and the Irish Catholic newspaper. The librarian in the library in Rathmore said that older people's taste in reading is eclectic. According to her, the types of books that are most popular among older library users are fiction, biography, history, DIY, gardening, books on computers and travel books. Reading was also a significant interest among the smaller sample in Rathbeg. However, older readers who cannot drive must depend on someone to get books for them in the library of an adjoining town. A mobile library service in Rathbeg has provided a rather erratic service in recent years. Those who attend the day hospital near Rathbeg can borrow books there.

While primarily an individual pastime, it is interesting that reading can also be enjoyed in a sociable and informal learning environ-

13 While word of mouth is still the most common way for people to hear about deaths in Rathbeg many people listen to the local radio for obituaries in case they miss hearing about a death and the arrangements for the wake and funeral.

ment. There is a Book club in Rathmore that attracts many retired people. They meet in the local library on a monthly basis and have a membership of approximately 20. A reading group is not as feasible in a small locality like Rathbeg as problems of transport and dispersed population may make it somewhat impractical. However, reading and story telling can be developed in settings where older people socialise and in residential and day care settings. One of ten projects selected from a newly established database of Healthy Ageing projects around the country was a Reading project attached to Wexford libraries.[14]

These leisure activities are easily accessed in one's own home and can be seen as enjoyable activities that act as a contrast to those compulsory and routine activities that must be undertaken for survival and maintenance. Therefore television viewing and reading will often be done in the evening when the daily chores are completed. However, while these activities can help to structure one's day and provide relaxation and reference points for understanding the wider society, they often involve little social interaction and provide little opportunity for expanding creativity or learning. However, we have seen that there are possibilities to develop these interests in an informal and enjoyable learning environment.

Outdoor physical activities

Sport, gardening and walking are common pastimes for older people. As we have seen outdoor sport was an activity that almost one quarter of the respondents (23.6 per cent) participated in. Indoor sport including bowls, snooker, darts and dominoes was played by a much smaller number, 6.7 per cent of respondents while swimming was enjoyed by seven respondents. Sport emerged as one of the major interests of

---

14 The Wexford project involves storytelling, critical reading of contemporary literature, reading-exchange, information technology training and literacy work. The programmes are run in a variety of settings including libraries, day centres and long-stay hospitals (O'Shea, 2003: 109).

respondents in the Rathmore survey, divided almost equally between those who participated in sport and those who followed sport as an interest. While soccer was the sport most frequently followed on television other sports followed included GAA football and hurling, golf, rugby, horse racing and tennis. While, interest in sport tends to be male dominated, several female respondents enjoyed a regular game of pitch-and-put and one respondent in Rathbeg spoke about how much enjoyment he and his wife got from watching rugby on television.

Gardening was very popular with two thirds of respondents reporting that gardening was one of their leisure activities and a significant number describing it as their major interest. While gardening may be seen as akin to home maintenance and as representing a rather 'privatised' lifestyle, we found that social interaction often occurred while people were in their gardens. Respondents spoke about talking to their neighbour 'over the garden wall' or 'the number of neighbours' who stopped to talk when they were working in their garden. Giving garden produce away to neighbours was common in Rathbeg. One respondent in Rathmore gave garden produce to the author when she was interviewing him. Having a well-kept garden may also be seen as a statement of pride in one's home and in one's locality.

Walking was a common pastime for our respondents. Many respondents walked regularly because of the benefits of physical exercise and if their health allowed they could continue this activity into their 80s and beyond. For some respondents walking was built into their daily routine in the course of going to the shops or the church. For others going for a walk was a weekend activity when they walked by the sea or in the park. One respondent, aged 87 years, who lived alone described his daily routine:

> I get the train at — station at 8.00 o'clock to Tara Street station. I walk to Gardiner Street church [for Mass] then to the Post Office in Lower Drumcondra Road and back at 11.30 – a two-mile walk.

A number of respondents said that if they were feeling down they went for a walk and felt better afterwards. These were people who lived alone and their experiences seem to concur with research on depression that shows that walking is an effective way to counteract 'down' moods. Moreover, this is an activity which people carry out in public, adding to the social landscape where they live.

A recent study of loneliness and isolation amongst older people in Ireland identified 'getting out and about' as an important aspect of older people's activity/organisational-related social networks (Treacy *et al.*, 2004). Participants in the study, which included both community and long-stay-care-dwelling older people, and irrespective of whether or not they had been out of their home for some time, had clear ideas about where they would like to go, for example, a visit to the seaside, a visit to a grave, attending Mass and many would have liked to do their own shopping. Participants identified such outings as important because it 'kept them in circulation' (Treacy *et al.*, 2004: 114).

In Rathbeg being able to be out and about, whether at Mass, in the pub or a daily walk was clearly important for our respondents. However, it was also important for older people to observe other older people socialising in public arenas. They would express satisfaction at seeing someone returning to Mass or resuming their daily walk again after a spell of illness or immobility. Walking was also a companionable activity that some respondents did with a spouse or friend while respondents in Rathmore spoke about how they enjoyed meeting and chatting to people on their walk to the shops. It was interesting that a number of respondents chose something simple like a regular walk as an activity they enjoyed regularly. Treacy *et al.* observed that there was not necessarily a need for great social contact on outings – the change of scenery seemed to be more important to many (Treacy *et al.*, 2004: 114). An informant in Rathbeg who also lived alone talked about how much she loved to observe the changing seasons on her daily walk.

The Health Promotion Strategy for Older People published in 1998 argues that physical fitness contributes to good health and offers opportunities for socialisation (Health Promotion Strategy, 1998). The Irish results of a European study that compared the participation rates in physical activity of the general population to that of the over-55s found that participation rates among the over-55s were very much lower in swimming and football and somewhat lower in cycling, dancing, keep fit and golf. On the other hand there was little difference in participation in walking while gardening was much more common among the over 55 group (Institute of European Food Studies, 1999).

Evidence from small locally based studies that consulted older people themselves suggested certain barriers to increased involvement in physical activities. In a study in Limerick older people said they would welcome the opportunity to be more physically active but many complained that community halls were often unavailable because of licencing and insurance problems. There was a lot of interest in countryside activities but concern was expressed about the quality of paths and walkways. There was also criticism of the lack of information relating to activities, events and courses (Limerick County Development Board Quality of Life survey, 2001). In Ballymun in Dublin issues identified by older people included lack of facilities, lack of programmes specific to the needs of older people, lack of transport and concerns about health/fitness and age (Ballymun Active Living Survey, 1999). The Ballymun study also highlighted very positive attitudes amongst older people about taking part in physical activity. A local group of older people organised nine activities every week in which 200 older people participated. According to information provided by this group the key to their success is that they keep their activities sociable, enjoyable and accessible.

Special interest activities

We found a wide range of special interest activities among our respondents in Rathmore and Rathbeg including creative activities. We found that art or craft work was an activity engaged in by approximately one fifth (19.4 per cent) of our respondents in Rathmore. While art may be thought of as an individual creative activity, art groups have become popular in both urban and rural areas attracting both older and middle aged adults. We found that a total of 18 people were involved in singing, music or dancing – singing in a church choir or group (3 per cent), dancing, usually ceilidhe or old time dancing (3.6 per cent) and playing a musical instrument (4.4 per cent). An interest in history was displayed through participation in local history societies, reading history or studying one's own family history. Creative writing was an interest that ten respondents enjoyed.[15] Knitting and sewing was another creative activity that women enjoyed while woodwork or DIY work was popular, mainly among men. Approximately three quarters of those who did DIY were men while one quarter were women.

While an activity such as knitting may be thought of as solitary, we spoke to a lady in Rathbeg who knits children's jumpers and sends them, through an intermediary, to children in developing countries. She showed the author a boxful of colourful jumpers which she produces at a rate of approximately one jumper every two days. She described her interest as something that kept her going since she is unable to go out and about much. What was also striking was her sense of connection with the children in countries such as South Africa and Romania who she believes benefit from wearing these warm jumpers even though she got no acknowledgement or no feedback about the children who wore them.

Activities such as art, music, creative writing and exploring history have a developmental structure and involve self-expression and reflection and can thus be seen as contributing to self-understanding and understanding the world in which respondents live. There is much interest in the qualities that define creativity and wisdom from psychological, philosophical and cultural perspectives. While there are ongoing debates about the effects that ageing has on cognitive processes of creativity and wisdom, it seems clear that gains in longevity, increased educational opportunities and more emphasis on lifelong learning have enabled older people to take up the challenge to develop themselves creatively (Moody, 1998: 411). Furthermore, research has suggested a link between participation in various leisure and educa-

15 One man in Rathmore has published many books on sport while another man in Rathbeg was writing his war memoirs. A number of respondents wrote poetry while one person told us that he contributes regularly to the magazine *Ireland's Own*.

tional activities and living healthier and longer. For example, research in Sweden found that attending cultural events, reading books and magazines and making music or singing in a choir had a positive effect on how long people live (Bygren, *et al.*, 1996).

Continuing education

The survey included specific questions on the respondents' experience of and interest in continuing education (Qs 8a, 8b and 8c). Only nine respondents said they were involved in an adult education class. Of these two were mature third level students while four were participating in art classes. This rate of adult education activity may appear low. However, Fahey and Russell found in their study that only 4.8 per cent of those aged 65–69 years had participated in education or training opportunities in the previous 12 months (Fahey and Russell, 2002: 81).

Approximately 15 per cent of the Rathmore respondents said they would enjoy the opportunity of further education courses. The most common preferences were for courses in art/crafts (seven respondents) and computers (seven respondents). Many of the respondents who said they did not wish for further educational opportunities added that they were 'too old'. Some said they would not have the energy and a number said that they might have been interested at an earlier stage of their lives. Comments that respondents made when answering questions about their formal education gave insights into their experiences at school many of which were negative: 'I got education outside of school not inside.' 'I was taken out of school each time my mother had a baby.' 'I would have loved to have done it, they were the bad old days … not much money and not much education.'

One respondent who spent a very short period in formal education, when asked if he would enjoy the opportunity for further education said: 'I'm well enough educated.'

However, there were positive comments too about the high standard of teaching in some subjects such as English when the respondents were at Primary school. A number of respondents who had gone back to education in later life spoke about how enjoyable it was and the tremendous self-satisfaction they got from it.

Studies of older people's leisure activities in other countries

A study that explored changes to the social and family networks of older people living in three urban areas in England examined the leisure interests of older people and compared them with leisure interests of a previous generation of older people. The authors commented on the extent to which activities have become either more solitary or couple-oriented when compared with the more communal activities that characterised multi-generational households and being 'vigorously at home in the streets' characteristic of household in the 1940s and 1950s (Phillipson *et al.*, 2001: 244). Reading and television in particular had become more salient than they were a generation ago. The authors suggest that these interests and the strong interest in gardening point to the increased significance of home for the respondents both from the aspect of the increased comforts in homes, the availability of home- based entertainment and the work associated with looking after the home and garden.

While there were important differences in the way in which questions related to leisure activities were asked in the English study and in the present study, some variation in leisure patterns in both countries is suggested. While reading, television and gardening emerged in both studies as significant activities, domestic and family interests appeared to be more significant in the English study. Participation in sport emerged as a major interest in the present study but did not feature among the first ten most popular activities in the English study. One reason for this may be the availability of golf courses in Ireland compared with London and environs. On the other hand walking was significant in the English sample and while walking was popular among the respondents in the present study it did not rank as a major interest. Another difference was that while knitting was popular in the English study, art was a more popular creative activity in the present study. In the present study current affairs, voluntary commitments and music were selected as major interests but they did not feature among the most popular activities selected by the English sample.

An EU sponsored comparative study of leisure activities of people aged 55 years and older in five EU member states (Ireland was not included) provided data on the variety of leisure activities engaged in, the cultural differences between countries and the factors that appeared to be related to participation in different types of leisure activities (The Mobilate Survey, 2000, European Commission). An interesting finding was that attachment to one's locality was an important factor related to participation in leisure in addition to resources of health, income and education. This finding resonates with a central idea developed in this study that participation in leisure promotes both a sense of self and of community. Activities were divided into two types, 'indoor activity' referred to activities at home while 'outdoor activities' referred to activities pursued outside the home and involved a much wider range of activities. Activities at home were the most popular such as watching television, receiving visitors, reading and talking on the phone. The most popular out-of-home activities were meeting friends, going for a walk and gardening. Other out- of- home activities were highly popular in some countries and had lower participation rates in others. These included religious activities, sporting activities, involvement in clubs, voluntary work and cultural events. The factor that was found to have the greatest influence on participation in both activities at home and out-of-home leisure activities was health. Factors related to the number of different out-of-home leisure activities pursued included, in addition to health, education, economic resources like income and car ownership, attachment to one's living area and age. Neither gender nor rural/urban environment influenced the variety of out-of-home leisure activities.

Social and personal significance of leisure activities

What do these leisure activities and major interests tell us about what our respondents perceive to be worthwhile activities? The activities appear to represent a mixture of individual hobbies, for example reading and gardening, and collective activities for example sport and voluntary work. They all involve some degree of endeavour, either physical, mental or both; however, while some are relatively undemanding such as television viewing, others, such as art, are developmental involving creative and reflective processes. While watching sport might not appear to involve much effort, there is considerable mental calculation and information to process to be up to date with league tables, scores, fixtures etc. In addition sport is an important source of conversation and contact between men. All activities appeared to represent respondents' pursuit of meaning in their lives whether exploring literature, doing a crossword, cultivating a garden, challenging oneself in sports prowess, going for a walk, enjoying watching excellence in sport, doing an art or craft activity or helping others.

However, almost one quarter of respondents said they did not have a major interest and many of what we would describe as more overtly purposive or developmental activities were minority interests. Furthermore, only half of the sample were involved regularly in at least one group. We pursue the theme of enjoyable and worthwhile social and leisure activities by examining responses to questions on the satisfaction derived from such activities.

Enjoyment from leisure activities

Respondents were asked about events and activities that they did for enjoyment or fun (Q. 34b). Over 70 per cent of Rathmore respondents said there were events that they did regularly for enjoyment or fun. However, over one quarter answered 'no' or 'not really' to this question. While up to three events were coded for each respondent most respondents mentioned just one. We have combined the answers into categories as shown in Table 4.5 below. The most commonly reported events enjoyed by 30.9 per cent of respondents were social occasions and meeting friends (this also includes going on outings, attending a social club and going to the pub). Outdoor physical activity, primarily sport but including walking, was chosen by 26.1 per cent of respondents. Special interest activities (including creative pursuits, going to cultural events, playing music, attending sporting events, bingo and bridge) were selected by 23 per cent of respondents. Occasions involving family and grandchildren were chosen by 10.3 per cent of respondents. Shopping or trips to town were mentioned by 7.8 per cent. Quiet and individual activities at home such as reading and television were chosen by 7.3 per cent of respondents.

Table 4.5: Types of event enjoyed regularly

Event	% of respondents
Social occasions and meeting friends	30.9
Sport and outdoor activities	26.1
Special interest activities	23.0
Family occasions	10.3
Shopping/going to town	7.8
Home-based individual activities	7.3
Other	4.8
n	120

n = 120 respondents who said there was an event they did regularly for enjoyment or fun.

The vast majority of events and activities enjoyed regularly were not home focused but belonged to the social world of friends, clubs and special interests. This confirmed the extent to which purposive leisure and leisure involving contact with others was a source of intrinsic satisfaction in life. However, a number of people said that nothing exciting happens in their locality, mentioning for example that a Senior Citizens' club had discontinued or that there were more social activities organised in their locality in the past.

When we crosstabulated the type of event that respondents chose first (Q. 34b) with their general enjoyment of life (Q. 38) we found that special interest activities, family occasions and sport were the activities most associated with enjoying life 'very much' Not having a regular enjoyable activity or choosing a home based activity were least likely to be associated with enjoying life 'very much'.

Having a major interest was also associated with enjoyment of life. Respondents who had a major interest were more likely to say they enjoyed life very much than respondents who did not have a major interest and the relationship was statistically significant (Table 4.6 below).

Table 4.6: Whether respondents have a major interest by enjoyment of life

	% of respondents			
Major interest	Enjoy life very much	Enjoy life quite a lot	Enjoy life sometimes	n
Has a major interest	56.7	37.0	6.3	127
Does not have a major interest	36.1	47.2	16.7	36
n	85	64	14	163

n = 163. Two missing values
Chi-square significance: p = 0.04

There was also a relationship between how respondents rated their health and whether they had a major interest – respondents who said their health was good or very good were more likely than those who said it was bad or very bad to have a major interest. This is not surprising given that health is a prerequisite for certain activities such as participating in sport. However, the relationship was not statistically significant suggesting that good health is not essential for the pursuit of a special interest. It was also interesting to note that being functionally active as measured by the number of daily tasks that the respondent could do was not significantly related to whether or not the respondent had a major interest. We will return later to the complex relationship between health and connectedness.

Class and gender differences in social activities

Many studies have noted class and gender differences in terms of participation in clubs and community social activities. It was not surprising that social class was significantly related to membership of groups. Respondents from white-collar classes were more likely to belong to a group than respondents from manual classes. They were also two and a half times more likely to say that they belonged to two or more groups and the relationship was statistically significant. It was interesting, however, that it was participation in church based groups where the biggest difference was found in participation patterns. White-collar respondents were far more likely to belong to church based groups than respondents from manual backgrounds. Of the 29 respondents who were involved regularly or occasionally with church based groups only four belonged to the manual socio-economic grouping. We will examine this in the next chapter in the context of religious practice. It was interesting that there was no statistical difference between male and female participation rates in relation to membership of sports clubs, charitable groups, community development groups or active retirement groups. Levels of membership in other groups such as Ladies' clubs, fund raising groups, music and drama groups and political parties were far too small to make statistical comparisons.

In relation to major interests white-collar respondents were more likely to have more than one major interest. Specific leisure activities that were more common among white-collar respondents included reading and crosswords and the relationship between both these activities and social class was statistically significant. Participating in sport was more common among white-collar respondents and going to the pub was more common among respondents in the manual classes, however, the differences were not statistically significant. While white-collar respondents were more likely than the manual classes to use the library the difference was not statistically significant. The location of the library in Rathmore means that older people living in its vicinity are more likely to use it. The librarian told us that some older people like to come in to sit down or to read the papers.

We have seen that the leisure interests of our respondents were patterned by gender, for example sport being largely a male interest and art/craft and bingo being largely female interests. The broader range of activities available in ARAs that includes for example bowls and art seemed to attract more men than did the activities of the more traditional Senior Citizens' clubs such as bingo, outings or music evenings.

The lack of participation by older men in community and health promoting activities is an issue of general concern among policy makers, service providers and community groups. A study of senior citizens in a traditional working class area of Dublin noted that men seemed to use the day centre to avail of the meal and then leave (O Sullivan, 2003). Estimations of how men spent their time involved activities such as going to the bookie's or the pub and walking the dog. Staff at another day centre nearby felt that men like to come along if there are other men present and that the less intrusive the environment the better as far as attracting men was concerned. They stressed the importance of 'normalising' care settings for older people In Chapter Nine we will return to the issue of differences in sociability patterns between men and women and between different social classes.

Summary

In this chapter we examined social participation through leisure activities and social clubs in Rathmore and Rathbeg. Social activities suggest many different orientations. We found that while many older people derived satisfaction from a variety of leisure interests and social commitments, there were significant numbers of older people who did not have major interests or who did not participate in communal leisure or in social clubs. Yet even leisure activities that might be thought of as passive, such as watching television, often involved a degree of sociability. We found that the leisure activities that were associated with satisfaction involved sociability with friends or family, shared interests, developmental processes and endeavour. The social relations of leisure are important since they often involve friendship with peers and engagement with others in the wider community. We argue that through communal leisure activities people create social worlds that make life meaningful and worthwhile. However, leisure interests can also be individual and solitary such as reading and help to create and maintain a rich and creative mental world for individuals. Sport and outdoor activities are important both for sociability and for maintaining health. Having good health is an important factor in the type of social and leisure activities pursued. Social class and gender are also differentiating factors.

The question of what is meaningful in the lives of older people is the central question we have pursued in this study. Our sense is that collective pursuit of interests and hobbies are hugely important in constituting a strong sense of self and community. However, to what extent is this bound up with broader attachments to the locality where one lives and to what extent does it represent a particular form of sociality? We explore the nature of relatedness in places and settings in Chapter Five and investigate perspectives on communal engagement. In Chapter Six we focus on involvement through religious practices, voluntary work and informal helping. The theme of how intertwined ideas about enjoyable and worthwhile social activities are with ideas about doing something for and with others will be explored.

Chapter Five
The Meaning of Place and Significant Communal Settings

Introduction

In this chapter we describe communal settings in Rathmore and Rathbeg where meaningful social engagement occurs and we investigate older people's attachments to the places where they live. We examine what place means in terms of sense of identity and belonging, and what relationships, rituals and practices are significant markers of attachment to place. We investigate the nature of relatedness in various settings and we develop the idea of enclaves of sociality where meaningful engagement with others is experienced. The role of intermediate groups in associational life and the different ways in which connectedness is created in localities are also considered. In order to empirically investigate the meaning of place we explore the likes and dislikes of respondents about their localities. We also examine respondents' ideas about the involvement of older people in their communities and consider the facilitators and barriers to greater social integration.

We begin the chapter by returning to a theme explored in Chapter Two in relation to the extent to which community involves face-to-face relationships and particularistic attachments or whether community can be based on identity and shared interests. We examine different perspectives on the idea of rootedness in a locality that is often seen as a feature of the older person's life. We refer to some recent evidence from Ireland on attachment to place in suburban locations – places often depicted as homogeneous and uniform and characterised by withdrawal from public life into the privacy of domestic life. We present data on what our respondents like and dislike about their locality, their sense of security and whether they desire to remain living in their present accommodation. We also present the interviewers' impressions of neighbourhoods in Rathmore obtained during the course of the study. We link the idea of attachment to place with associational life by describing settings and groups which are important social landmarks. We discuss how interactions in such settings help to create a sense of belonging and may facilitate authentic relationships and sense of solidarity with others. We discuss the respondents' views on how older people could be more involved in their communities. This gives insight into how communal engagement is seen from the individual older person's perspective. It also provides a good vantage point for viewing particular neighbourhoods with their unique configuration of amenities, services and groups.

Attachment to place and identity

We argued in Chapter Two that community should be understood as involving both place and type of relationship and that multidimensional relationships of the type described by Tonnies are well worth investigating and can indeed be found in contemporary societies. We carry this investigation further now by examining the significance of place in the lives of older people in Rathmore and Rathbeg. We ask to what extent locality and social settings of ordinary everyday living provide important contexts for the development of collective solidarities and transformative relationships.

While localities may of course be places where people simply reside for convenience or on a temporary basis, in childhood and older adulthood communities of place are seen as particularly important:

> Most people probably are to some extent, ineradicably constituted by their communities of place, the community defined by some or all of their family, neighbourhood, school, or church. It is noteworthy that dependent children, elderly persons, and all other individuals whose lives and well-being are at great risk, need

> the support of communities whose other members do not or cannot choose arbitrarily to leave (Friedman, 1992: 119).

In relation to the broader debate on how best to conceptualise community writers disagree on the extent to which community involves propinquity or whether community can be based on shared identity alone involving common interests (Gilleard and Higgs, 2005, Calhoun, 1994). In contrast to the importance that Gilleard and Higgs appear to attach to symbolic communities and communities of interest (Gilleard and Higgs:114–119), Calhoun rejects the idea that symbolic communities can offer support and belonging in the same way that a social grouping engaged in regular face-to-face interaction can (Calhoun: 111). In relation to older people's lives Gilleard and Higgs are sceptical about the idea of 'ageing in place' and attachment to place as a central dimension of community. They tend to frame their argument about community in later life in terms of two orientations – firstly, mobility through travel or migration, seen largely as positive, and secondly, staying put, seen as often resulting from lack of choice and lack of control over undesirable changes in long-established neighbourhoods:

> Assumptions about the dependency of older people on their neighbourhoods continue to be expressed, and much is still made of 'ageing in place'. While this rootedness is sometimes treated as a virtue … it can equally be viewed as a reflection of age-related disability – restricted mobility resulting from a lack of both symbolic and material capital (Gilleard and Higgs, 2005: 130).

However, in their criticism of the ideology of community Gilleard and Higgs tend to attribute negative experiences involving constraint and lack of individual freedom to non-voluntary attachments associated with place and other traditional sources of authority. Thus rootedness in a locality is seen as reducing individual autonomy and choice whereas 'modern retirees' who have financial means and who choose to live in sunny climes are seen as exemplars of 'modern living'. Indeed many theorists of civil society make too sharp a distinction between 'modern' or civil forms of community and 'traditional' forms. Most lives, and particularly the longer lives of older people contain a mix of relationships and commitments, some of which may be defined by 'old' communal ties to family, neighbourhood, church etc and some which may be defined by ties of friendship that are said to imply more choice and freedom. The lived experiences of people in Irish society, which arguably has more semi-traditional forms of relationships than England, suggest that we should not draw simplistic distinctions between 'voluntary' and 'involuntary' attachments:

> Irish evidence suggests that this distinction is greatly exaggerated. It challenges the assumption that we can easily differentiate between 'voluntary' and 'involuntary' forms of association, particularly at local level. It does not show that only 'voluntary' forms of association can be reflexive or transformative (Share, Tovey and Corcoran, 2007: 125).

While our lives are lived across a greatly enlarged set of physical spaces and while local societies can no longer be regarded as self-contained or isolated from the rest of society, it is not at all clear that attachment to local place has lost its significance for one's social identity. In Ireland there is evidence that people still seek and value an attachment to their locality. Furthermore, it can be argued that particularistic attachments to social group, neighbourhood or place are essential ingredients in understanding the nature of sociality. Living in a place carries an implication of having a stakeholding in a locality and associated with this an interest in the common way of life in that place. Edmondson has shown how in rural Ireland in particular property ownership has, or used to have, immense symbolic meaning in confirming the social identity of family members even when they have moved away from the family home:

> In rural societies especially, the 'home place' locates members of the family on a social map which helps to confer on them membership of a community; and the family home forms a focal point in which a couple's adult children, possibly scattered all over the globe, may gather (Edmondson, 2000: 22).

However, it is not only in rural localities that geographical place is an important mediator of meaning and sense of identity. Recent research found that a sense of attachment to locality is significant in shaping people's sense of belonging and sense of community in four suburban localities outside Dublin (Corcoran *et al.*, 2007). Against the main thrust of the literature the authors found that local residents were socially embedded in their respective localities. They felt attached to the place where they lived, they regarded the suburbs as good places to raise children and they derived sustenance from close relations with others, particularly those at similar stages of family formation:

> They are … inserted in networks of personal support which are locally centred and may participate in the public life of the locality in which they feel they have a stake. The sense of attachment to place depends to a large extent on the presence of friends, kin and good neighbours in the locality. The personal networks of social support are predominantly local and are mobilized in the daily lives of suburban families … it is largely through this embeddedness that the quality of life in suburbs is upheld: that residents feel positively about the place where they live; that families find the locality supportive; and that a wide range of services are provided and local concerns addressed (Corcoran, Gray, Peillon, 2007: 180).

Research from the US suggests that residents are more likely to describe their communities as 'family friendly' or 'caring' when there is a high proportion of families at a similar stage of the life cycle (Swisher, Sweet and Moen, 2004) or when there are high proportions of homemakers (Bould, 2003). However, the authors of the Irish study found differences in the social networks between some of the suburban areas that could not be attributed solely to variations at the level of individuals and households.

Likes and dislikes about their locality

We firstly asked our respondents what they liked and disliked about their locality (questions 29a and 29b). Their responses are shown at Tables 5.1 and 5.2 below.

Table 5.1: Aspects of locality that respondents liked

Aspect liked	% of respondents who mentioned this aspect
Convenient facilities and amenities	52.8
Nice neighbours/friendliness	39.6
Nice physical surroundings/ sea/park/etc	20.8
Sense of security/peaceful	15.1
Familiarity	11.9
Other aspect	9.4
Like generally/it's a good area	5.0
Privacy/non-interference by neighbours	4.4
Plenty of social activities	2.5

n = 159. Respondents could name any number of aspects they liked. Six respondents did not name anything they liked.

Table 5.2: Aspects of locality that respondents disliked

Aspect disliked	% of respondents who mentioned this aspect
Problems related to traffic, crossing the road and parked cars blocking access	25.9
Vandalism, crime, rowdiness	25.9
Noise, congestion, building work	16.7
Problems with apartment	11.1
Dislike neighbours/some people	9.3
Bad bus service	9.3
Untidiness/rubbish	9.3

n = 54. Respondents could name any number of aspects they disliked. Only one third of the sample named something they disliked.

What respondents liked about their area

All but 6 respondents (96 per cent of the sample) named at least one aspect they liked. A few respondents named up to four items and all aspects mentioned were coded. The positive aspect most frequently mentioned was *convenient facilities and amenities* – over half the respondents (52.8 per cent) mentioned this. They referred in particular to public transport, church and shops. Other facilities mentioned were schools and convenience to the airport. 'It has got everything I want and I can get to everything.' 'It is so convenient – chemist, grocery, solicitor, church, school – everything is within easy reach.' One man said that when he was working he had wanted to move to the countryside but he now appreciates how accessible amenities are in his area.

The second aspect most frequently mentioned (by 40 per cent of the respondents) was *friendliness and nice neighbours*. 'Neighbours ... we can totally depend on each other.' 'People in the area are so kind, considerate and helpful ... we've no problems.' 'The atmosphere when you go out ... the friendliness of the people ... always ready to give a helping hand.' This aspect is one that was a recurring theme throughout the study.

Having *nice physical surroundings* was mentioned by one fifth of the respondents – in particular the large well maintained public park nearby and being close to the sea and open spaces. 'There's plenty of open space, you don't feel hemmed in.' 'You can walk in the park, the beach, at the sea or the coast.'

Having access to a park or a green facilitated important social contact for a number of respondents. One man spoke about 'enjoying the football in the green at this time of year ... the juniors every Saturday and Sunday and often training in the evenings.' This man talked about meeting other people at these local matches and enjoying conversation with them.

A *sense of peace and security* was mentioned by 15 per cent of respondents. 'I like here cause we are safe here in the [a particular road].' 'It's nice and quiet.'

Familiarity of place was something that appealed to about 12 per cent of respondents. This was a sense of knowing one's surroundings and of having long-standing relationships with neighbours. But it was also the significance of memories and shared experiences associated with an earlier stage of adult life. People expressed this in different ways: 'I couldn't imagine [living] anywhere else … I would feel like a fish out of water … I like familiar surroundings.' 'We have all grown old together … reared our children together.' Other aspects mentioned by smaller numbers were non-interference by neighbours, the fact that is was a 'good area' and having plenty of social activities.

It is clear that facilities and amenities, both for practical day-to-day living and enjoyment of the natural environment are very important and valued by older people. Friendly and helpful neighbours are also highly valued and this suggests high levels of local trust. This provided further evidence that friendliness, kindness and practical social concern were experienced in Rathmore.

What respondents disliked about their area

When respondents were asked if there was anything they disliked about living in their area about one third said there was and two thirds said there was not. The two aspects most frequently disliked were *problems with traffic* and *crime/vandalism* – 14 respondents mentioned these respectively (8.5 per cent of sample).

The problems with traffic did not just apply to one area and included difficulties trying to cross the road or driving in and out of their houses due to parked cars causing obstruction. While some people attributed the traffic problems to the general increase in car ownership, a number of people complained about the response by the local authority and the Gardai to traffic problems. 'What they did to our road … they overdid the ramps.' 'We have an extreme amount of bother living in this area with drivers blocking the driveway … we cannot get in or out. I've been to the police and the Corporation … they do nothing … we're not listened to.' Concerns over crime referred to joyriding, vandalism, rowdiness associated with public houses in the area and break-ins.

Noise and congestion was a problem mentioned by nine respondents (5.5 per cent of the sample). Specific concerns included over-development and building work that caused annoyance.

Six respondents living in sheltered housing (one third of the sheltered housing respondents) mentioned *problems with their accommodation* most particularly the lack of a separate bedroom. Five respondents mentioned *problems with neighbours.* A few respondents spoke about 'new crowds' or 'a bad element' which had come to live in their area. Five respondents were *unhappy with the bus service* and five respondents were critical *of untidiness in their locality* citing graffiti and rubbish near the shopping centre.

The majority of respondents found nothing they disliked about their locality. However, problems with traffic caused annoyance to a significant number while crime and vandalism, especially related to anti-social behaviour such as joyriding and rowdiness, caused annoyance to a similar proportion particularly in one locality. A significant number of residents of sheltered housing expressed strong dissatisfaction with their bedsit accommodation.

Respondents' satisfaction with their homes and their sense of security

To get an indication of respondents' satisfaction with their housing and where they were living we asked if they wanted to change their living arrangements (Q. 36a and 36b). We found that only twenty respondents (12.1 per cent) wanted to change home. The most common desire was to move to a different area, usually somewhere quieter or outside Dublin (seven respondents), a bigger home (five respondents all of whom lived in sheltered housing bedsits), and a smaller house (four respondents). The remaining respondents wished for sheltered housing, a residential home, a home with no stairs and an apartment.

Respondents were also asked if they had a sense of safety or security in their area and in their own home. Approximately 90 per cent had a sense of safety in their own home while approximately 80 per cent had a sense of safety in their area. Only seven respondents said they did not have a sense of security in their own area while no respondent said they did not feel safe in their own home. However, many qualified their answer by referring to the threat or possibility of crime and most of the concerns seemed to be about break-ins. Some spoke about their homes having been burgled and felt secure only to some degree. Even respondents who felt generally safe in their home and in the area said that they had taken preventive measures by having an alarm and locks installed. Many respondents said that there was a 'Neighbourhood Watch' scheme in their area and this appeared to allay their concerns somewhat.

A number of people said they had been broken into more than once and one person reported having been broken into three times. One respondent said there was always a risk of being burgled when you were away. Another respondent said they always looked out for the homes of neighbours and kept an eye on who was calling. He commented that 'now with two people working it's a different ball-game.' Apart from concerns about house burglary some respondents had concerns about street crime and mentioned incidents that had happened to people they knew. In relation to problems of stolen cars, vandalism and gangs of young people hanging about, several respondents said that the proactive response of the Gardai contributed positively to their sense of security. Some residents of one sheltered housing complex said that the Gardai patrolled there regularly and one resident said 'there's no messing … the Gardai would sort it out.'

Anti-social behaviour

We noted that most of the concerns about anti-social behaviour above were associated with a predominantly working class locality; almost two thirds of those who mentioned crime resided in this particular locality. Specific concerns were expressed as follows: 'Stolen cars and people driving too fast … they've no respect … they try to frighten older people.' '[Named] Pub … there's fighting, it's noisy, attracts a bad crowd … there's drugs, littering, car damage.'

A needs analysis in this locality in Rathmore funded under a Community Development Programme (CDP) identified the following priorities for community initiatives:

- Improving the environment in the area (litter, graffiti especially around the shopping centre).

- Family support including more support for older people and carers, and
- Preventive measures to deal with anti-social behaviour involving alcohol, drugs and joy riding.[1]

In relation to older people specifically the report states:

> There is a sense that older people in the area are becoming more isolated. Problems of health and mobility affect elderly people, and also their carers. Accessible public transport services to health care facilities outside the area are poor and this limits people's access to hospital visits as patients or visitors. As people become more isolated in their homes the sense of anxiety about the outside world also tends to grow. There is an increasing perception among older people themselves that street violence is a problem and they feel vulnerable and afraid to go out.

Interviews with service providers indicated that there is little for older people in this area and that many have little motivation to get involved in social activities. One voluntary worker suggested that 'older people are looked after materially and individually but not communally.' Despite the presence of an active Ladies' club, bingo, a cup of tea after Mass one day a week and an Active Retirement group, many older people and perhaps men in particular do not seem to be well integrated in this particular community. We have already noted that a voluntary committee that organised social activities for older people in this area had more or less discontinued. However, we also noted the development of an initiative directed to lonely older people involving visitation by a parish group. This group was functioning well three years after its inception. However, there were more people who would like a visit than there were volunteers and the main organiser suggested that there was a lot of loneliness among older men in particular.[2] We will discuss the impact that both social class and gender have on communal engagement in Chapter Nine and we will discuss loneliness in Chapter Seven.

---

1 [Name of area] Social Needs Assessment Report, January, 2002.

2 Conversation with the organiser of the visitation group 17/10/05.

Impressions of Rathmore

We described the suburban features of Rathmore in Chapter One including the amenities, the facilities and the physical surroundings. It is clear that many older people value highly the convenience of having services and amenities close by them. Despite some negative experiences of anti-social behaviour, positive experiences of friendliness and neighbourliness were pervasive, enmeshed in commitments and relationships across many domains of everyday living. Both positive and negative neighbouring experiences are not always amenable to direct observation or measurement. The interviewers noted that different streets appeared to have their own 'atmosphere'. One interviewer described the difference between two streets which were not far apart where the people in one appeared somewhat mistrustful of each other, tending to make negative comments about some neighbours while in the other street people appeared more bonded. A second interviewer described the people in a particular street as being more 'guarded'. A third interviewer described the people at the top end of a particular street as more involved in the community than the people at the lower end. A fourth interviewer said that in one estate people were out and about more and greeted him while in a nearby street he observed that people seemed more distant from each other. These impressions were confirmed by some other informants in Rathmore. It was noted that two of the streets reported as not being friendly were long meandering streets which were car thoroughfares, there were no houses on the opposite side of the road and they were a greater distance from the shopping areas than other streets.

Public spaces such as parks, village squares, sports playing fields and green sitting areas are places where people can spontaneously meet and are seen as important for the creation and development of social capital. These are some of the enclaves where information is shared and people connect. Various studies have shown that the layout and design of residential areas and their impact on the functional distance between neighbours has a strong impact on patterns of social interaction and contact. For example, semi-detached houses with pairs of front doors facing each other are more likely to facilitate social contact than other arrangements (NESF, 2003: 108). However, it was not within the scope of the study to develop research questions about this more micro socio-geographic aspect of communality.

In the three sheltered housing complexes that constituted a specific research site there were clear differences in the amount of communal activities taking place and the involvement of residents in each of the three complexes. A Summer Activity programme that took place during the period of fieldwork in two of the complexes provided an opportunity for observations and insights. The cramped accommodation which many of the residents criticised would clearly make it difficult to have friends visit. However, the author was amazed at the high standard of tidiness and décor in each apartment. Each one had the appearance of a very comfortable and homely living room and the bedroom aspect was less obvious. There was great ingenuity in how space was utilised. One man's apartment had a study area with bookshelves, computer and a music area, all reflecting his wide range of interests and all contained within one room. The ingenuity with which some residents had transformed a very limited space was astonishing considering the size of each apartment.

Likes and dislikes of Rathbeg respondents

In Rathbeg respondents also said they liked the friendliness of people and the peace and tranquillity they experienced there. As regards dislikes several of the respondents referred to the lack of a community centre in Rathbeg. In Rathbeg at the time of the study many community events took place in the pubs which have limited space. However, a small resource centre suitable for meetings and small social gatherings was opened in 2005 attached to a new soccer club house. Among the findings of a local survey on what people believed should be priority areas for improvement were a day centre for older people and a community centre for all age groups. During the time of the study two local groups had plans to build a bigger community facility. In 2006 a new Social Activity Centre was opened under the aegis of a local charitable organisation that operates as both a day centre for older people and as a community centre where local groups can meet and social events can be held. This new centre has been a major social asset and after only one year in operation is well utilised. Further amenities desired include an indoor recreational facility for young people. In 2004 a new umbrella group representing local community groups was formed to develop enterprise and community amenities. The Rathbeg respondents did not appear to have major concerns about their safety or security.

The aspects of their locality enjoyed and appreciated by older people in Rathmore and Rathbeg resonate with the key themes identified by Corcoran *et al.* from the literature as being crucial in establishing a basis for attachment to place: the backdrop of the built and natural environment, the culture of place, elective belonging and associational life (Corcoran *et al.*, 2007: 180). The natural environment was appreciated as a place to walk and to enjoy nature while the built environment provided amenities such as churches, transport and shops that facilitated daily life, important rituals and access to services. The importance of culture of place is perhaps less expressible but can be seen in an appreciation of the history of a place and perhaps concerns that a place may be losing its character. The majority of older people in the study were long-term residents of their localities and some in Rathbeg had family ties going back over two centuries. They had strong ties with neighbours and friends and many had family connections. Even many of the residents of sheltered housing who were living for a much shorter time in their homes had formed enduring relationships with certain neighbours and felt attached to their locality. In Chapter Seven we will examine the nature of neighbourliness in Rathmore and Rathbeg in greater depth. We now go on to describe and discuss specific communal settings in both Rathmore and Rathbeg.

Settings for communal interaction

We propose to look in more detail at the nature of the informal interactions that occur in a selected number of social and leisure settings including the pub, bingo, clubs and special interest groups. We argue that the habitual social intercourse that occurs in public places and in intermediate groups helps to create social worlds of affirmation and meaning. The extent to which such groups promote connectedness as well as sense of self and sense of place will be emphasised.

Pubs

Over one third of the Rathmore respondents (36.4 per cent) go to the pub on some sort of regular basis. This is consistent with the findings of the Eurobarometer survey that reported that three out of ten older Irish people had visited a social club or bar in the previous week (Eurobarometer, 1993: 10). Of those who went to the pub regularly about 60 per cent were men and 40 per cent were women. In terms of frequency 39 per cent did so two or three times a week while 32 per cent did so once a week. People who go to the pub generally go to one in their locality, usually at the weekend with their spouse, other family members or friends. While many respondents, especially men, spoke about how they 'enjoyed going for a few drinks', only one respondent identified the pub as their major interest. Many women did not go to the pub at all or said they went occasionally or went for lunch. Some added that they preferred to go to the pub for a meal and a drink rather than for a drink only. Traditionally in Ireland women did not go to pubs (Curtin and Ryan, 1987). In Rathbeg older women, in contrast to older men, would never go to the pub unaccompanied. Older women may go to the pub with their families for special occasions and when family members are home on holidays. In Rathmore, however, while women went to the pub with their husbands, some went with an 'all-female' group of friends and relatives. One respondent stated: 'Sunday and Wednesday nights I go out with my sister and friends to the local pub.' Other female respondents said they went to the pub after an activity such as pitch-and-put.

The pub is an important social venue in Irish society and for older men who lived alone, and men who did not enjoy good health, the local pub was often their main social outlet. A conveniently located pub where they knew they would meet friends and to which they could walk or be driven was an important meeting place and something many of our respondents said they looked forward to. Some go a couple of nights usually at the weekend and meet friends there. A study of working class communities in an Irish town during the 1980s found that the local pub was central to the social life of the married working class male. Of 132 husbands surveyed 86 per cent went to the pub at least two nights of the weekend. Furthermore, half of them went to the pub on some week-nights (Curtin and Ryan, 1987: 140).

The present study suggests that the pub is also an important place for older men to go during the day. While many who go to the pub drink alcohol we also came across men who met their friends in the pub during the day for a cup of coffee and a chat. One pub in Rathmore offers reduced alcohol prices for pensioners and free newspapers. One respondent used the example of this pub to illustrate a need for a meeting place for older people and he was critical of the absence of a Parish Hall for this purpose. Another older informant who lives in a different area of Dublin suggested that the pub is a very important and underrated resource. He said that in his area the pub plays a large part in the lives of older men: 'It's not that they drink that much but it's warm and they can chat or not as they feel like.' He went on to opine that teetotallers have greater difficulty adjusting to retirement because they lack social ease that the pub engenders and they lack the outlet of the pub (Conversation on 3/3/04).

A visit to the pub referred to above in Rathmore on 10/5/04 confirmed that it is an important social venue for some older adults during the day. The author's informant who goes to that pub regularly during the day especially since he retired pointed out the different people whom he meets there. There was an older married couple who come in regularly and the husband reads the paper and has a drink while his wife goes to do the shopping in a nearby supermarket. They have a drink together when she returns. Another man had his dog with him who sat patiently beside him while he had drinks and read the paper. He claimed that the dog nudges him when it's time to go! A group of men came in from a regular game of snooker in a nearby leisure centre. Some of them are politically active and they were discussing the forthcoming local and European elections. A number of men sat at the bar. The author's informant said that their conversation usually ranges over sport, social welfare matters and 'gossip'. In addition to the older people a number of younger men were there, most of whom had finished early work shifts for example taxi driving. Many of the people there knew each other and there was conversation between the different groups. Just as the author left a game of dominoes was starting.

In most areas in Ireland there is a choice of pubs, however, many middle aged and older people choose to frequent one pub. In urban areas where the choice is wider the decision about which pub to frequent on a regular basis reveals something about the social relations in an area and individuals' perceptions of the social groupings in their area. One man who goes to a pub on the periphery of Rathmore three times a week spoke about his reasons for not frequenting a more local pub in a nearby sports club:

> The clubs here were a bit rough with rows and drink … no way am I getting into it … familiarity breeds contempt. The [named] club was too much of a clique, they were too much with each other and they all rowed. They wanted me to leave the [named] pub [his regular pub] and come there but I said 'no'.

Sometimes people choose to go to a pub outside their own area because of a position they hold in the community. We noted from our interviews with priests and some parish workers that they generally do not go to the pub in their Parish, preferring a drink in a pub or hotel where they are not known.

In Rathbeg during holiday periods older people are brought to the pub by family members home on a visit. There they meet and greet people that they would know but whom they would see infrequently. However the lack of an alternative community venue for older people in Rathbeg who never go to the pub represented a serious gap in community amenities and was referred to by many of the respondents in Rathbeg. A local hotel, now demolished, served as an important social venue in Rathbeg for over one hundred years. People of all ages used to frequent it for a sociable drink, a meal or to attend a function or meeting. It has been described rather 'tongue in cheek' in local conversation as 'Rathbeg's sitting room'! Some older people who were happy to go there for a drink do not feel comfortable in the pubs. One reason is that the pubs tend to be smaller, less comfortable and more crowded at holiday periods. However, the tightening up of the drink driving laws in Ireland in 2005 may be having an impact on the pub-going patterns of older men in rural areas in particular.[3]

Bingo

Bingo was an activity that one tenth of the respondents in Rathmore took part in and bingo was referred to as a major interest by three persons. It is primarily a female interest – we have noted that of the 17 respondents who participated in bingo, 16 were women. Bingo takes place in small groupings such as Senior Citizen's clubs and day centres. Bingo also takes place on a larger scale in community halls, sports clubs and function rooms of hotels and is used for fundraising by sports clubs. Many older people, who would not attend bingo as an event in itself, enjoy bingo as part of their club's social activities which also involves meeting their friends and having a cup of tea. For others bingo is a more serious interest and they travel to the larger bingo sessions outside their own locality several times a week. The importance of bingo as a social outlet, could be seen in the lives of some respondents. A widow who lives alone and who said that she suffers a bit with her nerves said that bingo was her major interest and that she looked forward to chatting to people at the bingo. In Rathbeg bingo takes place on a fairly large scale in different localities four or five times a week and some people go to several bingo sessions every week.

Bingo is often represented as a pastime of no relevance and is frequently dismissed as an uninteresting form of entertainment. This is a misrepresentation. Bingo is a very important activity that enhances social connectedness in the lives of many older people in Rathmore and Rathbeg. Playing bingo requires a degree of concentration and reasonable hearing and sight. If the location is convenient or a lift is available people can continue to play bingo into late old age. In Rathmore a regular bingo session in one of the community rooms attached to a sheltered housing complex attracts a number of ladies who are in

3 'Future of rural pubs under threat, group warns', Article by Sean Mc Connell in *Irish Times*, 16th May, 2007.

their 90s and one lady attends who uses a walker. In Rathbeg people tend to travel together to regular bingo sessions in a shared car or minibus. News is exchanged and then relayed later to family and other acquaintances. For some older people the bingo outing is their main contact with others in their community. It provides a sense of belonging in the same way as other forms of communal leisure. It makes an individual part of the loop of gossip and local news that is important for feelings of belonging in a community. The author has observed this herself among family members and relatives. In Rathbeg a minute's silence is always observed when a participant has died. This is very much appreciated by the person's family and friends.

The prospect of winning is of course what adds an element of excitement to bingo and 'Any luck at the bingo?' is a question asked frequently of bingo goers as news is relayed of the previous night's winners along with any other news exchanged among bingo goers of events such as illness, death, accidents, local crimes and so on. It is apparent that people are very pleased if they win something, even the small prizes that are given out at the small-scale bingo sessions. However, at the larger bingo sessions, attendance is always higher when extra prizes are offered for example turkeys and hampers at Christmas or occasionally when the prizes are given in sterling instead of euros, a practice that happens in Rathbeg. Another aspect of bingo observed by the author was the help given to people who had difficulty marking their card due perhaps to deteriorating eyesight. This is done in a quiet way that does not embarrass the person being helped and enables that person to continue to participate in bingo even when faculties of sight and hearing have deteriorated.

Bingo does not appeal, however, to the majority of older people and a particular drawback is that it can dominate the programme of day centres and may deter certain groups of older people from going to a local centre. A study in Dublin's South inner city found that bingo was the most popular activity that took place in all day centres in the surrounding area. It was played predominantly by women and staff and volunteers reported fierce opposition from the older people themselves against any attempt to introduce other activities at the expense of bingo (O'Sullivan, 2003: 70–71). Yet, concern was expressed that the centre in question was underused. The policy implications of taking into account the preferences of current day centre users, while also adopting a developmental approach through provision of a wider range of activities, and trying to attract potential users such as other senior citizens, active retirement groups and men, are discussed in the report (O'Sullivan, 2003: 52).

Special interest groups: Art groups

In Chapter Four we discussed the wide range of special interest activities among the respondents, the growing popularity of art groups for example, and the developmental aspects of such activities. We will describe two special interest group settings that we found in both Rathmore and Rathbeg – art and history groups. Art has become a particularly popular hobby in urban and rural localities and many respondents spoke about how absorbing they found art and how they looked forward to it. We found that Art was usually practised in a group context organised through classes or ARAs.

In Rathmore an art group has been in operation for the past twelve years meeting weekly in the community room of one of the sheltered housing complexes. On the day that the author visited (18/05/04) there were five older women painting water colours under the guidance of a teacher who was herself in the older age group. She said that the numbers attending are higher in the winter months and that attendance has ranged from 23 participants to the rather smaller number present on that day. Some of the participants were preparing work for an exhibition in the local library the following month. The art group is under the aegis of a Senior Citizens' club and many of the women were former volunteers in the meals on wheels service. They have presented their work at an annual art exhibition organised by the Federation of Active Retirement Associations (FARA) in the Writer's Museum in Dublin. They also exhibit at the local library along with other groups, including residents of a Retirement home in the locality. Two of our female respondents in Rathmore ran art classes in their own home. One 78 year old lady, a widow, gave evening classes in an Art studio in her own home and these were attended by people in the area.

An art group was set up in Rathbeg in 2000 and has approximately 12 members, most of whom are in the middle and older age group. The members pay for the teacher and for the use of the premises each week. They take it in turns to supply tea and biscuits. They have held an annual Art exhibition for the past several years to display and sell their work and it has become an enjoyable local social event. One aspect of the art group in Rathbeg is that it conveys a sense of place. Many of the paintings in the exhibition are of local beauty spots and places that are recognised. The significance of participating in an art group seems to be both the social aspect and the sense of achievement generated. One respondent in Rathbeg, who had four paintings on display at the art exhibition in summer 2004, when asked what it had meant for her to get involved in art, said:

> It has meant a big change in my life. I never did any art at school, I never painted, I never drew a line. I used to always say I couldn't draw a straight line, I still can't. But I went to the art class just to get away, because I was looking after (name of relative), it was becoming very stressful at the time, and somebody suggested: 'why don't you join the art group'? But I went and I enjoyed it. We had an exhibition and at the first exhibition I sold a picture. And that really set me going, it was a great boost to my morale. Anyway, I keep it up. I enjoy it. I'm not an artist nor never will, but I enjoy dabbling and I enjoy the company. I suppose it's more for company than anything else. It means going out once a week for 3 hours and we have great fun (Interview with respondent in Rathbeg on 17/03/01).

The social aspect of art was also evident from the words of a respondent in Rathmore, a widow who gives art classes in her own home and who said that since she took up teaching art she never feels lonely. We have already discussed the growing interest in developing opportunities for creativity including creative writing, painting and craft activity in day centres and residential settings for older people in Chapter Three.

Historical societies

Historical societies are groups that tend to attract older people as members. The role of older people as repositories of history and transmitters of shared and collective identity to a younger generation has been discussed by writers such as Erikson and Laslett:

> As collectors of time and preservers of memory, those healthy elders who have survived into a reasonably fit old age have time on their side – time that is to be dispensed wisely and creatively, usually in the form of stories, to those younger ones who will one day follow in their footsteps (Erikson *et al.*, 1986: 331).

Laslett discussed many potential social benefits of the Third Age including that of older people acting as trustees for the future in relation to preservation of material and cultural heritage:

> This sense in older people of an indefinite social future, one that does not confine itself to their own descendants, immediate or distant, is accompanied as must be expected, by a strong sense of the past, the personal and collective past (Laslett, 1996: 194).

There is an active Historical society based in Rathmore, many of whose members are in the older age group. They meet twice monthly and organise regular talks open to the public on aspects of local history. They have produced two publications on aspects of local history – a Census of the area from 1901 and 1911 and a study of its major historical buildings.

In Rathbeg a Historical society was established in 1996 with initial membership of six persons. One of the aims of the society is to collect and record local history in various forms, including photographs, written and oral material. Many of the older people in the locality have contributed to their work by giving photographs and oral recollections of their younger days. The society has produced several publications in recent years documenting by photographic and oral recollections aspects of the history of the locality. Commemoration of a major national historical event that occurred in Rathbeg was being planned during the period of the study.[4] The involvement of older people in this project and other heritage projects is evidence of both their capacity to plan for the future and of their appreciation of the rich cultural heritage that they wish to preserve and pass on to future generations. The secretary of the Rathbeg Historical society spoke about the first project that she had undertaken involving the publication of the photographic record. When she began working on the book, she said she had been working and living away from the locality for many years and wasn't really involved in the community. She said she was amazed at the interest that the older people she approached had in the project and the extent to which people wanted to reminisce. She spent hours talking to people in their homes because they wanted to tell their stories, to identify who was in photographs and to ensure that the secretary knew what the connections were between people:

> You never went anywhere without getting all the history. It made me feel part of the community. They weren't just interested in their own history but in the people who were there with them and the way of life (Interview with Secretary of Historical society on 20/08/04).

She also spoke about 'a richness' in their stories and 'the descriptive way they created a picture of a way of life by recalling where they were when they heard something of interest, who was with them, their turn of phrase'. This account illustrates the appreciation that older people in Rathbeg had for their cultural heritage and the value they attached to preserving and passing it on to future generations. It also suggests that a group involved in a collaborative community project such as this can help to create a sense of community. In terms of understanding sociality it was also significant that for those undertaking the project a sense of a way of life was evoked not by people's recollections of events per se but by their recollections of their interactions with those who shared their work and life space.

4 The success of the commemoration of this national historical event in 2007 was in considerable measure due to the work of the Historical society.

Habitual interactions in places

The significance of cultural practices involving greeting and conversing with others in one's locality whom one does not know intimately was emphasised throughout the study and will be discussed further in Chapter Nine. Many daily interactions in neighbourhoods that might not be regarded as purposive or developmental activities expressed connectedness, for example watching football matches in the local park, having a chat with someone when you are in your garden, going to the pub or meeting people on the way to the shops or at Mass. While greeting others whom one does not know particularly well or having a chat with a fellow spectator at a local match do not involve active engagement or reciprocal relationships, we would argue that such forms of social intercourse are important for generating a sense of solidarity and belonging.

These types of interactions can be seen as a form of generalised friendship. Edmondson had previously described a form of 'distributed intimacy' in her ethnographic work in the West of Ireland. Edmondson suggested that communicative forms associated with traditional societies mean that emotional satisfaction may be derived from daily contacts spread throughout the locality in which one lives and does not necessarily rest only on intimate and personal relationships (Edmondson, 2001: 68).

We have described the habitual interactions, rituals, and cultural context which create a sense of self and a sense of place. The activities and interactions described show how one's life interweaves with the lives of others who share life space and locality and gives people a sense of belonging. Furthermore, these settings provide important contexts to form authentic relationships. The quality of associational life through informal and formal interactions at local level are also crucial to attachment to place. We will discuss the significance of community and voluntary associations in Chapter Six. We wish to examine a further aspect of place attachment and communal engagement, that of older people's perspectives on engagement in associational life.

Ideas on how older people could be more involved in their community

The survey respondents were asked whether they had any particular ideas on how older people could be more involved in their community. Over half of them (55.2 per cent) had ideas while 44.8 per cent had not. However, not having ideas on this question did not in any way imply that these respondents did not regard many aspects of community life in a positive light. Had they been prompted or given examples of what might be entailed by 'more involvement in the community' it is likely that many more of them would have given positive responses in the same way as those who had ideas. However, since we wished to find out how the respondents perceived associational life in their localities and what 'more involvement in the community' might mean from their particular perspective we did not provide any prompts.

Many of those who had ideas explained them very clearly and referred to their own experiences and own locality. While many expressed one central idea, others had more than one idea and we coded all ideas mentioned. The individual comments were particularly insightful into what respondents perceived to be the value of community engagement, the possibilities for satisfaction in older life, the difficulties older people experienced and the impediments to greater social and cultural integration. Some of the comments from those who said 'no' also provided insights into their experiences and perceptions such as: 'I don't tend to focus on the community.' 'No, there's enough around doing that.' 'No, I'm happy with my life.' 'Not really, I'm not a joiner … it's not a fair question.' Table 5.3 shows the different ideas grouped according to different themes.

Table 5.3: Respondents' ideas on how older people could be more involved in their communities

Ideas	Number of responses
There are adequate services and activities	24
Mentioned barriers to community involvement	24
There should be more centres and clubs	23
It depends on individual tastes and preferences	17
They should remain active and involved	17
Other	15
Gave examples of community involvement	13
Should be more involvement between younger and older people	6
Mentioned political aspect	5

n = 91 respondents who expressed an idea. More than one idea could be expressed and all ideas were recorded.

Twenty four respondents said that there were adequate services and activities available and referred to those that were enjoyable or helpful for older people. The types of activities mentioned included Active Retirement groups, Senior Citizens' groups, Ladies' club, bingo and meals service: 'A lot of people lose out by not going out and joining things, there are things and one thing leads to another … there's enough to suit all tastes … Active retirement are very good.' 'In this community there are a lot of facilities, transport arranged … day in All Hallows with meals … there's a great sense of togetherness.'

Many spoke about groups attached to church: 'They have a Widow's Mass and bowling … a lot go to that.' Many people said that in their area older people were well looked after: 'There are plenty of good services in this area for older people, Old Folk's etc.'

In contrast with these comments were views expressed by 23 respondents that there should be more organised activities and social centres for older people. Many residents of sheltered housing believed the community room attached to each complex should be utilised more:

> Open the place up [community room] … different games going on like they have in England … somewhere you can go during the day … darts and dominoes … the pub is the only place here to go and chat and sit down and have meals. There's nothing in the courts for anyone … we're behind things compared to England.

Others believed that the church should be used more:

> If you'd your parish hall, free newspapers … the church is locked after 10 o'clock … it would be a sin not to use it for something after people paying … more things might develop if it was opened as a centre. The [name of pub] pub does that, reduced prices for drinks, and newspapers free … a lot of older people amble down there.

People spoke about the need for a community hall: 'There should be a community hall for daily meetings with tea/newspapers away from family for all religions.' Some respondents proposed that schools be utilised more since they are under utilised in the evenings and during the summer.

A number of respondents believed that older people living alone needed more support:

> Old people living on their own would need support … the Legion [Legion of Mary] used to do more … some people would like to meet others while getting a meal instead of the meal being delivered … they can be lonely … there should be more social work support.

The need to have more day care centres was mentioned: 'There should be more opportunity to be involved in day centres like [name of day centre] ... you get Mass, meals … it's very sociable.'

One respondent referred to the disbandment of the Senior Citizens' club and said there was now nowhere to go.

A large number of respondents (24) spoke about barriers and inhibitors to older people being involved in things including reclusiveness, shyness and laziness. One respondent said: 'It's up to yourself to get out [but] I'm kind of shy … I'd prefer if someone would come and take me out to something.'

Respondents in the sheltered housing courts who were active in organising activities or who participated in activities frequently mentioned the difficulty in getting people involved. One respondent said:

> Some of them don't go out … that's their life, they're too into themselves. Some didn't get married … a pattern has built up … [name of person who organises activities] and I are going to try to get them out … they're very negative, never positive about anything. They're all suspicious of each other, they become odd from living alone, they suffer from their nerves, they have their hang-ups. You have to respect their privacy … we might try an outing at Christmas and in the summer. We will all help if someone is sick … we all watch out for one another.

That sporting/leisure activities can become competitive and prevent people from continuing their involvement was mentioned by one respondent. A female respondent said that she and her husband used to play bowls but were isolated from it because of his bad eyesight.

Social activities attached to the church did not necessarily appeal to everybody. One female respondent said: 'I'm not involved … a lot of people are not inclined to join church groups ….'

Many respondents (17) referred to individual tastes and preferences determining what older people were involved in. Some emphasised the importance of respecting that individuality, others suggested that some people are more self-sufficient than others and don't need communal activities: 'Everybody seems to do what they are capable of and what they are interested in.' 'Everyone to their own … some people are more independent and enjoy their own company.' 'People are as they are.'

Another set of responses (17) suggested that it was important to remain active and involved in the community. Some respondents suggested doing voluntary work or being helpful to neighbours: 'I do believe voluntary work should be taken up by older people … people who don't have much to do in the day.'

A resident from one of the courts said: 'People should try to get involved in things … you're living here, you have to consider helping people and people helping themselves.'

Another respondent said: 'Older people are inclined to withdraw which they should never do.'

Many people gave examples of community involvement including specific community events (13) and some mentioned particular individuals whom they regarded as making a major contribution to their community. Many residents from the sheltered housing courts mentioned a community festival that was held or about to be held during the summer. They expressed the view that it would get people out and mixing. One female respondent spoke about the Festival afterwards:

> We had something for every day for the Festival week … we started and finished with an Ecumenical service … dances at night and a keyboard player … strawberries and cream. We're looking forward to next year.

This same respondent paid tribute to the work of one individual in particular: 'There's one man in the complex and there's not enough hours in the day for all he does.' She went on to itemise some of the work he did.

The contribution that older people make through the church was acknowledged: 'In the church the older people are doing all the work.' 'I already mentioned our own community development … I'd like to see more people getting involved in that … occasionally we get one or two others.'

The importance of older people having contact with younger people or of younger people being helpful to older people was a theme mentioned by six respondents. One aspect of this was that it was psychologically healthy to mix with younger people. One lady who was widowed when her children were young said:

> Mixing with younger people if you can. I don't agree with older people together … you're just looking at the next stage … some are decrepit. I never joined the Widows' Association because looking at others like you was not what I wanted.

Another respondent believed that younger people should do more for older people especially those living alone and not robust: 'Younger people should step in the shoes, help the people who have no one, people should come in and do some gardening and try to make their lives a little easier.'

The idea of secondary school students visiting older people whom they know during their Transition year was suggested by one respondent. Another idea mentioned by three or four respondents was that older people could help younger people by teaching them the skills they had such as cooking and sewing, by talking to them about the past and helping them to appreciate the difficulties that life can bring and by helping those with a drugs problem.

A political aspect was mentioned by five respondents including the suggestion that older people were too passive and that they should have an input in to health and social services planning: 'Older people could have more of a say and take part in the operation of the health services and how they are run'.

> They should be a bit more demanding … they kept the country ticking over until the Celtic Tiger arrived … they should work together, band together coming up to election time. Some older people are too passive … Irish people are passive in general.

Finally there was a group of responses that were categorised under 'other' because they ranged over different themes. Two responses were particularly interesting because they reflected broader social processes of change particularly as they have affected the role of men: 'I don't know … I think there is too much "we'll all get together and be happy" – we were reared to be independent in my day.' 'Men are reluctant to be too involved with the community … it's too dangerous … that is a general feeling … the wife is involved.'

The perception that people were more independent when the first respondent was young suggests interesting questions about definitions of masculinity and changing understandings about individual autonomy and the role of the state. The second comment appears to refer to the climate of revelations about child sexual abuse by males in positions of authority that pervaded Irish society during the 1990s and on in to this century.

It was noteworthy that, regardless of whether or not respondents perceived that there were adequate facilities and supports for older people in their own locality, there was a common central idea that community involvement was life enhancing for older people through for example clubs and leisure/ social activities, volunteering and social contact through visiting and social services support. It was striking how many respondents appeared to appreciate the work of those voluntary and community workers who tried to provide social activities and encourage more participation. Many respondents also recognised the impediments to greater social participation, especially for those older people who were not 'joiners' or who had become set in their ways or somewhat reclusive. There were a number of interesting suggestions about intergenerational involvement and one or two critical perspectives on older people socialising exclusively with each other.

Place and associational life

The multidimensional quality of relationships found in places may depend on some degree of continuity of residence. In the present study over 80 per cent of the sample of older adults in Rathmore had lived in their homes for at least 30 years and the shared experiences of families and individuals at similar stages of the life cycle had contributed to warm social relations. Familiarity with place involving long-standing relationships with neighbours and shared memories was a theme mentioned quite often by residents. A local Parish Priest in reflecting on his understanding of the parish as a community said:

> They've [parishioners] a regard for the short history of this Parish. It's only 35 years old. But still they would talk about the Priests who were here and what they did, and the time when the schools were bursting at the seams, and there were Priests and people working together to form school football clubs and all that. All of that will come out at funeral time … And elderly people even who would come to a funeral wouldn't know the Christian name of the neighbour who died. That was all out of respect, a great regard and respect for each other (Interview with a Parish Priest in Rathmore, 2nd May 2001).

We would argue that localities can best be thought of as places peopled by others who become pointers on our social map who help us to locate ourselves. An important function of place may be to facilitate shared memories that help us to see our lives as narrative and as interweaving with other lives. Thus the importance of place seems strongly related to seeking and establishing authentic relationships based on the elements of shared experiences, common identity, interdependencies and friendship. Among the older adults in both Rathmore and Rathbeg social interactions with long-standing acquaintances – be they relatives, friends or neighbours – and rituals and practices associated with familiarity of place and shared memories appeared enormously important.

To stress the importance of place is not, however, to suggest that social relations have a unique quality in a particular place or to attempt to draw definitive geographical boundaries around places. Network analysis makes a strong case for taking a broader perspective on relationships and including both proximate and distanced relationships. Bulmer reminds us that physical proximity involves both subjective and objective conceptions of one's neighbourhood (Bulmer, 1996: 21). Moreover, communities of place do not, in fact, have completely uniform sets of practices and values – as we have seen they will always to some extent be overlapping and plural. Nonetheless we would suggest that community as both a set of reciprocal interdependencies and as an experience of authentic human relationships is experienced in places.

Chapter Six
Helping, Volunteering and Religious Practice

Introduction

We have demonstrated the extent to which the social and leisure activities of older people in Rathmore and Rathbeg provide opportunities for intrinsic satisfaction and mutuality of experiences and in these ways help to develop creativity and social connectedness. Possibilities for older adults to live full lives were strongly related to opportunities for satisfying and developmental activities in meaningful interaction with others. We have also discussed the communal contexts associated with place which help to create a sense of belonging and the possibility of experiencing authentic relationships.

In this chapter we will examine how older people experience and express connectedness through religion, volunteering and informal helping activities. We wished to know the extent to which religion was an important source of meaning for our respondents and to investigate the formal and informal interactions involved in religious practice. The helping activities that older people are involved in, both through formal voluntary work and informal help and support to family, friends and neighbours, was also a central question for investigation. The value frameworks involved in helping activities and the extent to which these activities represented 'vital involvement' are important themes in this chapter. A shared moral sense, even within heterogeneous and pluralist societies, is important in establishing identity and mutual obligation. However, the extent to which and the ways in which common moral frameworks develop amidst diversity and difference, characteristics of both rural and urban communities, are difficult questions that we can only hope to shed some light on.

Questions for consideration included: Were forms of sociality involving helping others or volunteering different from those involving leisure? Did an active pursuit of hobbies and interests suggest a focus on enjoying oneself and imply a less socially committed orientation? Or did both types of activities add up to a productive or fulfilled life for individual older people? To what extent did the social worlds associated with leisure, voluntary work and religious practice overlap or represent separate domains of interest? How do these forms of sociality help to define the values in civil society? Are there particular types of social and community groupings that help sustain individual and community well-being as people age? Through trying to answer these questions we hoped to develop further insights into how older people experienced communality and what forms of relatedness were associated with their perceptions of a good life.

We begin the chapter by a brief examination of the concept of social capital which purports to explain the nature of social relatedness and the conditions and benefits of associational life for individuals and societies. We discuss religious practice in Ireland and refer to theoretical approaches to the study of religion and spirituality. We examine religious practices among our respondents and focus on religion as an important meaning system. The role that older people play in church groups and the forms of social interaction involved in religious practices are also examined. We examine the extent and type of voluntary work undertaken by our respondents and the meaning and motivations involved. We refer to the wider context of voluntary work in Ireland and different orientations towards voluntary/community work. We investigate the value orientations implied and the links between volunteering and religious practice. The extent to which volunteering provides intrinsic satisfaction is also discussed. In the final part of the chapter we examine informal helping that older people are involved in either as givers or receivers and we highlight the considerable amount of positive neighbouring in Rathmore and Rathbeg.

Volunteering, religion and social connectedness

In examining connectedness and personal and social well-being among our respondents we need to look more closely at what people do within different social networks and how their activities relate to public as well as private goods. The concept of social capital and the debate on what it is and how it can be measured have emerged as central ideas in understanding the social relatedness of human beings, their well-being and the forms of relatedness that conduce to a good life and a good society. Helping others which is strongly encouraged by philanthropical and religious traditions is seen as an indicator of social connectedness but the nature and extent of helping may differentiate traditional from contemporary communities. According to Robert Putnam the principle of generalised reciprocity is the touchstone of social capital (Putnam, 2000: 134). Philip Abrams linked positive neighbouring with generalised reciprocity while Martin Bulmer made the further suggestion that generalised reciprocity among neighbours could develop into commitment and friendship. Putnam argued that both altruism and honesty are important diagnostic signs of social capital and are bolstered by dense networks of social exchange. Thus, he argued that our willingness to help others and to trust others develops through social networks that foster norms of reciprocity and attention to others' welfare (Putnam, 2000: 117; 136). However, greater mobility and choice associated with processes of industrialisation and global capitalism mean that the neighbourhood may be less significant as a locus of social interaction and social support. Furthermore, strong neighbouring such as that found in traditional societies may be associated with collective deprivation and may not be as relevant in more affluent societies.

Social capital

Social capital is generally taken to refer to the ability of actors to secure benefits by virtue of membership in social networks and other social structures (Portes, 1998: 6). Social capital is viewed as a social resource in itself additional to that of physical and human capital; it is seen to be accumulated over time and can be drawn upon and used in a way that produces personal, economic and social gain. Interest in social capital from a public policy point of view arises because of the idea that social ties and associated norms of reciprocity are potentially productive. We examine the concept of social capital to see if it can help us to answer a question posed by Graham Crow (2004: 16): 'What it is about how people come together that enables them collectively to make the most of their individual skills and attributes?' More particularly can the concept of social capital help us to understand the nature of sociality in Rathmore and Rathbeg?

There are limitations and difficulties with the concept of social capital. Firstly, the concept of social capital is morally neutral. Secondly, there is a difference between social capital in terms of structure and content. It has been suggested that in the debate in the US and in Britain the distinctions between social capital and older policy concepts such as community have been blurred. This has led to an assumption that social capital and social cohesion are the same thing (Field *et al.*, 2000: 262). Putnam argued that the vibrancy of associational life is positively correlated with the stocks of social capital a society possesses and he suggested that social capital has declined in the United States due to a fall-off in associational activity (Putnam, 2000). However, the number of organisations that people belong to tell us little about the strength of social capital unless we are told, a) what people actually do as members, and b) how far this relates to public as well as private goods (Field *et al.*, 2000: 28). Social capital can exclude groups as well as create connections and understandings among and between groups. For example not all associations involve the same degree of face-to-face interaction or networking or political awareness or bridging across groups.

Edwards and Foley argue that it is the socio-cultural component of social capital and not the fact of membership of voluntary associations that provides the context within which it acquires meaning and becomes available to individuals and groups. Thus, their understanding of social capital is very context dependent, involving inter-subjective understandings that may produce either trust within imagined communities or distrust across them (Edwards and Foley, 1997: 671). Therefore, we must not assume that all aspects of communality are positive or conducive to the good society. Enforced communality can militate against personal development and more diverse social relations may help people to gain access to wider networks thus enhancing their life chances and opportunities. Granovetter's concept of the 'strength of weak ties' contradicts the assumption that only voluntary networks composed of ties that are both dense and strong create social capital (Granovetter, 1973).

We must ask, therefore, what specific aspects of community conduce to personal and social well-being. Furthermore, there are limits to the extent to which volunteering, for example, is a panacea for problems and ills in society. While volunteering may under certain conditions represent an important social resource, voluntarism alone is unlikely to meet complex social needs and is no substitute for well run state services, especially for incapacitated older people, some of whom may have limited kin or friendship networks, who need reliable forms of support that are acceptable to them and that promote their autonomy as individual citizens.

At the same time we do take seriously the ideas of writers such as Putnam who argue that doing things with others, whether through informal sociability or membership of formal clubs, is a powerful stimulus to altruism and helping activities. Putnam suggested that while doing good *for* other people may be laudable it is not part of the definition of social capital. It is, he suggests, when we do things *with* others that we reinforce bonds within the community (Putnam, 2000: 116–117). Furthermore, we were interested in exploring the extent to which activities such as volunteering or helping others involve communicating an agreed set of expectations about appropriate standards of behaviour that may be associated with high levels of trust and a sense of collective responsibility. The idea that intermediate groups provide settings where positive forms of sociality can be modelled for others is worth exploring further.

It has been argued that Faith communities in which people worship together are the single most important repository of social capital in America. Church going and saying that religion is very important are especially strong predictors of volunteering and philanthropy (Putnam, 2000: 66–67). Voluntary work is frequently undertaken in a religious context as part of one's religious duties and frequently through the institutional structures of a church. To what extent are the Christian churches in Ireland important sources of interpersonal and social connectedness? Is it the case, as Putnam suggests, that the social ties embodied in religious communities are as important as religious beliefs per se in accounting for volunteerism and philanthropy?

Given that religious practices have been particularly strongly embedded in Irish culture, churchgoing and other religious practices may not per se be associated with voluntarism in a way that might be manifested in more pluralistic and secular societies. It has been argued that the ideological controls exercised by the Catholic church in Ireland made religious practice and adherence to Church teaching part of the taken-for-granted world and strongly embedded in cultural practices. Inglis argued that the devotional and orthodox religiosity of traditional Catholics may be seen in terms of a religious habitus that was inherited and embodied deep within their social being (Inglis, 1998: 18). Therefore, religious practice may not be associated to the same extent with other, perhaps newer forms of civic involvement such as community development approaches that often emerge outside the ambit of the Churches.

Religious practice in Ireland

Ireland is predominantly a Catholic country. However, as is the case in many advanced industrial societies in the West, traditional religious practice has declined rapidly in Ireland particularly in more recent decades. While the Church was clearly important to the older people in this study, this contrasted somewhat with overall trends of decline in religious practice in Ireland. According to surveys showing trends on weekly Mass attendance, there was a decline from 87 per cent of Catholics to around 65 per cent of Catholics who attend weekly Mass from 1981 to 1998 (Fahey, 2002). This pattern has accelerated with a 2002 IMS survey showing weekly Mass attendance at 48 per cent by September 2002 (*Irish Times*, 20th September, 2002). In terms of variation by age and geographical location the earlier surveys found that Mass attendance varied from 94 per cent of older people in one Diocese outside the Capital to 50 per cent of all Dubliners to 41 per cent among all 18–24 year olds in the country (Fahey, 2002). There are different explanations for the decline in religious practice in Ireland and Western Europe generally and the meaning of such decline. Inglis analysed the religious field in Ireland as it relates to other social fields (Inglis, 1998). He conceptualised the declining position of the Catholic Church as a dilution of the moral influence of the Church which symbolically dominated Irish life in a pervasive way. Where, for example, Inglis contends that in the past 'attending Mass on Sundays was a key strategy in maintaining social capital' the situation now is that:

> Irish Catholics can attain status, honour and respect, and they can attain political, economic and social power, without having to have Catholic religious capital or to have their forms of capital symbolically legitimated by the church (Inglis, 1998: 67; 205–206).

Davie provides a different perspective on religion in Western Europe. She relates the decline in institutional participation and traditional indices of religious belief to parallel declines in confidence and participation in public institutions and activities in other spheres – government, electoral processes, trades unions and even sport (Davie, 2000). While such trends may indicate a decline in communal life, there are counter-trends that suggest that religion continues to play a continuing role in locating the individual in relation to a tradition. An interesting line of theoretical inquiry is Brown's (2001) work on the connection between discursive religion as a social force and religion's ability at the personal level to enable people to make sense of their lives in narrative terms.

This links with the work of Alistair MacIntyre who sees the normative disintegration of modern societies as connected with this loss of narrative (MacIntyre, 1985). MacIntyre develops an account of the relationship between 'practices', virtues and their social context and consequences. He emphasises the need for daily interaction in communities and the moral force of daily habits and practices. MacIntyre's ideas about the kind of civil society groupings that foster civic virtues led him to extol small communities and religious groups in particular. Thus, MacIntyre's communitarian critique of modern society concludes that virtues cannot easily be cultivated in contemporary Western societies because of the individualism and acquisitiveness of the modern economic order. However, it can be argued that secular humanist traditions (such as Fabianism) and Christian traditions share core values, that they are both concerned with the creation of a moral community and that the concept of virtue is important in both traditions. Furthermore, there may be more traces of communality in contemporary life than MacIntyre allows.

Religious practice in Rathmore and Rathbeg

Reflecting the religious composition of the population in Ireland[1] the overwhelming majority of the sample in Rathmore were members of the Catholic Church, a handful were members of the Church of Ireland and a small percentage were not members of any church or were agnostic. Religious practice was clearly important to the vast majority of older people in the study. Levels of regular church attendance were very high; almost one third of respondents attended Mass daily while

1 The 2002 Census classified the population of Ireland by religious denomination as follows: Catholic – 88.4 per cent, Protestant denominations (Church of Ireland being the largest) – 3.8 per cent, Christian (unspecified) – 0.6 per cent, other religions – 1.6 per cent, no religion/not stated – 5.6 per cent (2002 Census of Population, CSO).

about 40 per cent attended at least two or three times a week. A further 40 per cent attended Mass or religious service weekly, making a cumulative total of almost 80 per cent who attended Mass or religious service at least once every week.

As in Rathmore the majority religion in Rathbeg is Catholicism. There are two other active churches – Church of Ireland and Presbyterian.[2] There is a strong adherence to institutional religious practice among both the Catholic and Protestant communities. Many older people are daily Mass goers. Respondents who could no longer attend Mass due to disability or infirmity said they missed going but if they attended a day centre Mass is often included in the programme. In addition many Parishes have parish radio that relays Mass to people who are sick or housebound.

Salience of and expression of religious belief

It was beyond the scope of this study to investigate religious or spiritual beliefs in any great depth. Nonetheless, respondents did give us insights into their beliefs and the importance of their religion in coping with life and as a source of existential meaning. Respondents were asked whether religion or spirituality was important to them and any additional comments they made were also recorded. Almost 70 per cent said that religion or spirituality was very important, 17.6 per cent said that it was fairly important while just under ten per cent said it was not very important. A few respondents said that spirituality more so than religion was important.

Not surprisingly church attendance was closely correlated with the importance of religion or spirituality in the lives of the respondents and the relationship was statistically significant. For example of those who attended daily Mass 98 per cent said that religion or spirituality

2 While membership of these Protestant churches is proportionately small, both churches have regular church services and members are active in church related groups. The Protestant church is strong in Donegal compared with its overall numerical strength in the Republic of Ireland making up 10.8 per cent of the population of the county. (*Protestants in Community Life*, 2001: 2).

was very important to them while 63.5 per cent of those who attended weekly Mass said it was very important. While religion and spirituality were not treated as separate subjects in the questionnaire, it did not appear as though spirituality was separate from religion in the lives of older people in the study. Nonetheless, religious observance was not necessarily a reliable indicator of the salience of religion or spirituality in the lives of the respondents. For example, over 40 per cent of those who attended Mass infrequently or never said that religion or spirituality was very important to them. On the other hand 11.1 per cent of those who attended weekly Mass said that religion or spirituality was not very important to them. This highlights the complexities of personal meaning systems involving elements of religion and spirituality and the inadequacies of survey data to elicit subjective and intersubjective experiences. While spirituality has emerged as a separate subject of investigation from religion in the literature, this distinction was not emphasised in the study (see for example Pargament, 1997, Dillon and Wink, 2007).

There were different approaches evident in relation to religion and spirituality, from the devotional and pious to a more questioning reflective stance, and in some cases a critical and disenchanted attitude towards the Catholic Church. However, prayer seemed important to many; regular personal prayer, prayer in coping with difficulties and prayer through Mass going and church attendance. Many respondents commented on the importance of prayer in their lives: 'I wouldn't be able to get through my life without my prayers.' 'All you have got is the prayer.' 'I pray a lot especially if I want something.' 'I couldn't do without my few prayers – I rely on God, Our Lady, St Martin.'

One respondent spoke about a particularly difficult time in his life when prayer helped him: 'When [a difficult predicament happened] I got down on my knees … I could actually feel the grace … I just thanked God for my blessings … I got up a changed man.'

Many older people say set prayers including the Rosary and prayers they learned as children and they include in these prayers all their family and people they know who are ill or who have personal predicaments. While high levels of Mass going and prayer indicate strong belief among the older age group, it is also clear that these religious practices were developed from childhood and were supported by the rhythms of communal life and institutional forms of authority that do not prevail to the same extent in contemporary Irish society.

At the same time the trends toward more private spirituality and secularism can be contrasted with the continuing popularity of forms of prayer such as Novenas. Novenas remain popular and have been incorporated into liturgies that attract high levels of 'voluntary' church attendance. For example a Novena of Grace that is held in a church in Rathmore every March attracts a packed congregation on eight consecutive evenings. However, such forms of religious practice appeal more to older adults and most of the congregation at the Novena of Grace are in the over 50 age bracket. It was observed that the Parish priest urges those attending to make an effort to invite a neighbour in to their house for a cup of tea during the week of the Novena and on the last evening there is a social gathering in the Parish Hall. We referred in Chapter Two to a significant religious ritual, the annual graveyard Mass, that takes place in many towns and villages in Ireland.

For many older people religion is interwoven into their lives and can be observed in their use of language, for example when speaking about the dead. One of the interviewees in Rathbeg spoke about deceased members of the community known to both the respondent and the author and he invariably added after each deceased person's name: 'God be good to him and all those belonging to him' in a natural way that suggested that faith and religion were not compartmentalised in his life. We have referred to the embeddedness of religion in everyday life in Irish society in the past. We see evidence here of the 'religious habitus' that Inglis described in cultural practices that involved the maintenance of a holistic lifestyle in which older people's religious beliefs and practices were not separate from the rest of their daily lives (Inglis, 1998: 94).

Religion and social interaction

The importance of social ties embodied in religious communities has been recognised. Older people are particularly active in church groups. In addition religious practices are part of daily and weekly routines and create opportunities for many social interactions as well as legitimising various forms of encounter and interventions. For many people who went to Mass daily or a few times a week Mass was built into the structure of their day and had a social aspect to it. One man who attended Mass two or three times a week said: 'It gets you on an early start and you meet people.' Another respondent who attended daily Mass said:

> I get up at 7.00 every morning. I know the people who live around [name of area]. After Mass every morning we have a chat. I meet fellas, hear the news, what's going on … they're well up to date on everything. One man I speak to – I don't know his name – I chat about Gaelic football with a neighbour on the [name of road] and I don't know his name (Interview July 02).

The significance of cultural practices involving greeting and conversing with others in one's locality whom one does not know intimately was emphasised throughout the study. In a diverse society where privacy is also valued it may be argued that membership of a common moral community, in this case membership of the Catholic church, makes intrusion in the form of greeting and chatting after Mass 'legitimate'.

Involvement in church groups

The five Catholic Parishes in Rathmore have active participation by lay people and older people are particularly active in parish work. Indeed many of the respondents said that it was older people who do the bulk of work in Parishes and the author found much evidence of this. We have noted that church groups were the most common group that our respondents were involved in, with 15.2 per cent of the Rathmore sample helping in their church in some capacity. The types of work that older adults did in their parishes included being members of Parish Core Groups or Parish Councils, serving as Ministers of the Word (reading the scriptures at Mass) or Ministers of the Eucharist (giving out the Eucharist at Mass), cleaning the church, helping to prepare the Liturgy, helping with the church collection, singing in the church choir, helping with fund raising and so on. Parish centres are an important focus for social activities in many communities.[3] One parish directory lists 44 ministries within the Parish, not necessarily all connected with church work, but all use the premises of the Parish Centre. Church related groups listed in the Directory include a Liturgy group, Folk group and St Vincent de Paul Society, while other groups listed include a Widow's group, Bowling club, Ladies' club and Residents' Association.[4] However, this parish in Rathmore would be regarded as a particularly active one.

The church based groups active in Rathbeg include Lay groups such as the St Vincent de Paul Society, Pioneer Association, the Legion of Mary and Ministries of the Word and the Eucharist. Older people are well represented in the composition of these groups. In the Church of Ireland the Vestry which is made up mainly of older people runs the church's affairs locally. The Reader explained that everybody has to be involved because their community is so small. The Reader who is herself in the older age group has responsibility for conducting the service including preaching the homily in the absence of the Minister, a role she described as 'very fulfilling and challenging.'

A very striking aspect of involvement in church based groups was how differentiated it was by social class. Table 6.1 below shows a strong statistical correlation between involvement in church based groups and belonging to a white-collar background. This raises important theoretical questions as well as questions for church leaders about lay involvement in church groups. Does this apparent dominance of church groups by the middle classes represent a middle class cultural form associated with regularity, discipline and rearing of children? Or does it suggest a degree of social confidence among middle class people where they operate with greater social ease in certain institutions or settings such as the Church? Within the social capital literature Hall has drawn attention to a widening of the difference between the levels of social capital available to the working classes

3 Of the five Catholic parishes in Rathmore in 2003 only one had a parish centre. A parish centre was near completion in a second parish, the third parish had plans to build a community/parish centre, a fourth had meeting rooms attached to the church and the fifth had a parish office only.

4 A Rathmore Parish directory, 2003. The parish centre was opened in 1997.

and the middle classes arguing that 'the forms of civic involvement available to the least privileged are more fragile in the face of secular trends' (Hall, 1999: 458). From his analysis he concludes that the level of social capital seems to be most strongly sustained by the middle class.

Table 6.1: Involvement with church groups by social class

	% of respondents		
	Social class		
Involvement with church groups	White collar	Manual	n
No involvement/previous involvement only	52.2	47.8	136
Regular/occasional involvement	89.7	10.3	29
n	97	68	165

Chi-square significance: p = 0.0

Change and conflict in religious practice

High levels of church attendance and cultural adherence to institutional religion do not of course imply consensus or lack of conflict in relation to matters of public religious practices. While Churches provide an important mediating structure and at parish level an associational infrastructure that fosters solidarities, many people choose to participate in social networks unrelated to churches. Some people have negative feelings towards the church because of bad experiences of church personnel or may feel excluded because of the Church's stance on matters related to marriage or sexuality.

An example of conflict arose in relation to saying Mass in a day centre opened in 1997 in the environs of Rathbeg. Because of the high proportion of Catholics in Ireland and because so many older people are practising and committed Catholics, an assumption can be made that Mass is an activity that is desired and appropriate in day centres. By an informal understanding with most of the day centre users and the staff, the local priest started to say Mass on the two days that the centre was open. One of the residents of the small sheltered housing unit attached to the day centre said he would not come to the day centre any more as he did not want Mass and felt it was not appropriate given that people from other denominations also attended the centre. The Committee discussed the problem and decided that the practice of regular Mass there should stop except on special occasions and by prior arrangement. However, the decision caused a split in the Committee and considerable personal distress to individual committee members who found it difficult to be on the opposite side of the priest in a rural community. The Committee had to request the intervention of the local Bishop to help resolve the dispute.

Changes in religious practice are clearly in evidence in Rathmore and Rathbeg. It has been suggested that many European societies are now witnessing the emergence of what Bellah termed 'civil religion' where people seek and use religious values and symbols without the social commitment involved (Bellah, 1967). Such practices are certainly not evident among the older people in this study. Nonetheless, the high rates of churchgoing found among older adults in Rathmore and Rathbeg do not apply to the same extent among younger age groups and are unlikely to persist into the future. One informant stated: 'Are we in danger of being too complacent or too comfortable because we go to church in big numbers? ... and what of the future?'

Further changes in religious practices are likely due to a shortage of priests in Catholic parishes which may make it a necessity for lay people to take on work in Parishes traditionally carried out by the priest. For example one informant, a retired person and an active member of the Parish group, described how, in the absence of the priest, he was asked to call to the families of two deceased parishioners to discuss arrangements for the funerals including selecting the Scripture readings. He described how anxious he was about doing this as he was unsure how the families would react to a lay person carrying out this function that was traditionally the role of the priest. A Parish priest in Rathmore speculated that after his retirement the Parish would not have a full-time priest. He was philosophical about these changes and positive about the increased lay participation that it would imply, stating: 'It is sad that the church is getting smaller but not sad that it is losing its power'.

Voluntary work undertaken by older people in Rathmore

We found that almost 30 per cent of our Rathmore sample did some form of voluntary work. A sizeable number of the sample, 11.5 per cent, did two types of voluntary work while 4.8 per cent of the respondents did three types of voluntary work. When we combined all types of voluntary work undertaken and included more than one voluntary commitment we found that charitable work, community development work and church work were the most common forms of voluntary work undertaken by older people in Rathmore – charitable work was mentioned by 18 people in total (10.9 per cent), community development work by 17 respondents (10.3 per cent) and church work by 16 (9.6 per cent). Sport related voluntary work and informal voluntary work were undertaken by eight and five respondents respectively, while we found one or two respondents in our sample who were involved in the running of a Ladies' club, a Senior Citizens' club, a Music and Drama group and an Active Retirement group. It must be emphasised that these patterns refer only to voluntary work undertaken in such groups and not to membership or participation in the activities of these groups.

Examples of 'charitable voluntary work' undertaken by our respondents included the St Vincent de Paul Society, meals on wheels provision, visiting housebound or lonely older people, counselling bereaved people, helping in a voluntary nursing home and helping in an organisation for people with disabilities. In terms of level of commitment, we found, for example, that many of the volunteers in the Meals service and members of St Vincent de Paul had given huge amounts of time to this work over a number of years. We have already noted the variety of Church work that older people do. 'Community development work' includes Residents' Associations and Credit Unions.

The level of commitment of some members of Residents' associations was impressive, particularly in the sheltered housing complexes.

In line with other Irish studies on volunteering there was no significant difference in the proportions of older men and women who volunteered (*Tipping the Balance*, 2001: 16, 17; Fahey and Russell, 2001: 77). One man in the present study said he believed that men were more reluctant to volunteer now because of the climate of allegations of sexual abuse in Ireland in recent years. In line with international trends and consistent with the other Irish studies a positive association was found between social class and volunteering (Powell and Guerin, 1997:131; Fahey and Russell, 2001:79). The data indicates that approximately 70 per cent of respondents who volunteered belonged to a white-collar background while 30 per cent belonged to a manual background. Age was also strongly correlated with volunteering. Almost three quarters of volunteers were under 75 years while approximately one quarter were over 75 years.

The rate of volunteering among older people in Rathmore appears somewhat higher than average trends as measured in other Irish studies. Data from three surveys on volunteering carried our during the 1990s suggested a national rate of volunteering of 33 per cent for all age groups and 25 per cent for those aged 60 or more (*Tipping the Balance*, 2001: 16–17). However, survey data collected during 2001 from a national sample of people aged between 55 and 69 indicated that one third of all respondents were involved in some sort of organised voluntary work (Fahey and Russell, 2001: 77). When account is taken of the age group of the Rathmore sample, over 65, it suggests a significant input by the older age group into voluntary and community activities in Rathmore.

While the authors of *Tipping the Balance* raised the possibility that the potential pool of volunteers among those who are retired is not being sufficiently tapped at present, Fahey and Russell drew opposite conclusions. They found that there had been an increase in voluntary activity among the age group 55 to 69. Furthermore, from the point of view of encouraging unpaid helping activity, it was those who were partially retired and who did some paid work who had by far the highest rates of voluntary work. They conclude that the partially retired have that apposite combination of time and resources which enable them to contribute heavily to voluntary work.

Voluntary work in Rathbeg

We found in our more rural community of Rathbeg that the types of voluntary work undertaken by older people there were similar to Rathmore. In all three Christian churches in Rathbeg older people help with various Church tasks including cleaning, maintaining church grounds, assisting with the liturgy and helping with fund raising. Visiting people who are lonely, usually older people at home or those in hospital or residential care, is done in an organised way through the St Vincent de Paul Society and the Legion of Mary.[5] An interesting form of charitable voluntary work was carried out by one of our respondents, a retired British army officer who, for many years, looked after the welfare needs of war veterans and their dependents in Donegal who had fought in the British army. This involved visiting war veterans and their widows all over the County and ensuring that they and their families were receiving their entitlements. This respondent was awarded a CBE[6] for his work in 2003.

Older people are active in special interest and community development groups. A local Historical society has produced publications about local history and played a key role in organising commemorative activities for a major Historical event. An Art group provides a creative outlet for local amateur artists and organises a local exhibition every year. A local Enterprise group acts as an umbrella group for all organisations and groups involved in local economic and social development. Older people are also active in the Care of the Aged Committees in Donegal which we described in Chapter Three. Considerable voluntary effort is involved in maintaining these and other community and social groups and this work promotes an important form of connectedness both for the volunteers and the wider community. These groups, with financial backing from public bodies, have helped to pro-

---

5 Organisation of Catholic laity.

6 British honours award of Commander of the British Empire.

vide marine and angling facilities, environmental projects, heritage projects, football facilities and local festivals.

Community development approaches

Community development activities in Rathmore and Rathbeg are concerned with developing some aspect of the locality, either from an enterprise perspective, to improve the general environment or to achieve greater social inclusion. This would include the work of Residents' Associations, community development projects and local enterprise groups and it is important to point out that community development activities are not confined to the interests of any one age group. An important supportive framework for community development activity has been provided by the state through a Community Development Programme (CDP) funded by the Department of Community, Rural and Gaeltacht Affairs since 1990. At the time of the study there were two projects funded under the CDP scheme in Rathmore and one in the environs of Rathbeg. The community development projects supported by the state are mainly in areas of social disadvantage. Traditionally community development approaches have not been used in work with older people, related perhaps to a perception that older people need to have things done for them, not with them or even by them. However, an innovative community action project, 'Voice of Older People' which was established in County Donegal in 2000 has acted as a catalyst for the formation of Active Age groups and the empowerment and enablement of older people.

Voice of Older People project

The project 'Voice of Older People' was funded by the North Western Health Board in 2001 as a 3 year pilot project. The aim was to develop a representative and recognised forum of and for older people in Donegal to enable them to collectively tackle health and social issues that affect their lives. Local groups have been formed in different areas of the County that are run by older people themselves. Their aims are to identify the needs and concerns of older people, to agree tasks and priorities for each group, to influence decision makers, to engage in advocacy and information sharing and to engage in fun activities. Each group sends representatives to a Forum to coordinate strategies at the County level where there is direct engagement with policy-making groups and statutory bodies. Priorities identified by the local groups to date include the development of a telephone networking project to access information, link people with their local community, engender a sense of security and reduce social isolation. Other issues identified include intergenerational work, elder abuse and housing options for older people. The project team organised two open days in Donegal in two different parts of the County which attracted 390 and 270 participants respectively.

This project was selected as an example of best practice in an evaluation study of Healthy Ageing projects in Ireland which reported many positive outcomes from the project (O'Shea, 2003: 99). The personal and social benefits accruing from the Voice of Older People project suggest that such innovative projects could act as a catalyst for greater participation of older people in their communities.[7]

In recent years Active Age groups have developed in parts of Donegal and at the time of the study there were 15 Active Age groups in Donegal. Each group is autonomous and decides on a programme of activities that interest members. For example one group meet for a swim and for lunch once a week and organise trips to the theatre. Another group organises yoga classes and gatherings for spiritual reflection. What may be significant about all such groups is that, irrespective of their goals and specific interests, they involve doing things with other people rather than for other people. This means that the individuals involved both get to know each other and develop shared understandings of the good of the wider group/community. This does not imply that there are no disagreements on priorities or approaches.

7 Specific actions that were being taken at the time of the study arising from the project included development of a cross border senior help line, community development/leadership training courses, collection of stories on islands in North West Donegal and development of a post retirement pool of people to work as handymen in a social economy project.

But it does suggest a form of social connectedness that promotes both personal satisfaction and social concern.

Volunteering, values and social participation

We wish to pursue the meaning of voluntary work in the lives of older volunteers and the value frameworks implicit in various types of voluntary work. A criticism frequently made of voluntary organisations is that they have had a weak commitment to participation by service users (Faughnan and Kelleher, 1993). We noted the tendency for some traditional clubs for older people to be closely identified with the founders and principal organisers and we found that this sometimes created an impression of the group being a 'closed group' and unresponsive to change. This aspect combined with a high average age among committee members may militate against continuity and renewal of such groups. One observer[8] who is involved in the community in another capacity opined that 'the volunteers themselves got too old ... there was nobody younger trained in ... people hold on to positions for too long.' While many older people prefer a quieter and more passive role in a social group, principles of participatory democracy are more evident in some groups than others.

On a visit to a Senior Citizens' club in Rathmore the author observed a clear demarcation between the organisers and the participants despite the obvious warmth and friendliness between them. While bingo was being played the committee members brought tea to the bingo participants, then went into the kitchen to take their own tea together. While there are clearly functional reasons for this sort of interaction and thcir taking tea together in a sociable way demonstrated the friendship between the committee members, it is unclear how directly compatible this is with all the ideals of participative democracy. This is not meant as a criticism but as an observation of differences in how clubs were traditionally run and newer models such as ARAs.

8 Interview with voluntary worker in Rathmore 12/10/03.

Churchgoing and volunteering

We have suggested that there are strong links between religious orientation and volunteering. Using British social attitude surveys Gill found a statistically significant correlation between participation in voluntary service groups and church going. Thus, 27 per cent of members of voluntary service groups were weekly churchgoers compared with 11 per cent of the total sample. A further 42 per cent attended church at least monthly compared with 18 per cent of the total sample and only one-third said they never attended church compared with two-thirds of the total (Gill, 1999: 107–8). Furthermore, the forms of voluntary activity most closely related to membership of a religious group were services, residents/tenants associations, women's groups, political party, pensioners' and trade union groups. In contrast, sports and social club membership was least closely related to religious group membership. Gill concludes that 'the more regularly individuals go to church the more likely they are to share and practise distinctively Christian virtues. While these virtues are not absent from the broader society, they are found disproportionately amongst churchgoers and especially amongst the most regular churchgoers' (Gill, 1999: 200). Herbert argues that the evidence provided by Gill in relation to the cultivation of Christian virtues through regular church attendance correlates with the civic virtues discussed by MacIntyre and can be translated into the more community-building-oriented parts of the voluntary sector. He concludes that religious organisations foster the kind of participatory citizenship on which an active civil society depends (Herbert, 2003: 92).

Bulmer in his review of Philip Abrams' work on neighbours, neighbouring and informal care suggested that a major basis of informal social care is religion: 'About three-fifths of Good Neighbour schemes could equally well be described as Good Christian schemes … people with a strong sense of religious commitment are disproportionately active' (Bulmer, 1986: 234). A somewhat different interpretation of how membership of a church may be conducive to generalised reciprocity was suggested by Abrams. He found that significant numbers of voluntary helpers are enabled to bridge the distance between themselves and others by the belief that common membership of some moral community, typically a church, makes 'intrusion' legitimate (Bulmer, 1986: 114). On the other hand Bulmer suggests that more research is needed to establish exactly why religion is an enabling factor in involvement in informal care. However, the wider range of groups referred to in Gill's study suggests that helping activities were not confined to people of one's own church and involved broad other-regarding values.

In the present study there was a statistically significant correlation between churchgoing and volunteering but it was not as strong or as clear cut as in Gill's study. Table 6.2 below indicates that respondents who went to church two or three times a week were the group most likely to volunteer while those who went on a daily basis or once a week were equally likely to volunteer. Respondents who were irregular or non-churchgoers were less likely to volunteer while not surprisingly no respondent who participated in church services through radio or television was a volunteer. How can we interpret this? In Gill's British study where churchgoing is not as culturally embedded, belonging to and practising one's religion is perhaps more purposive. The same values and dispositions are often involved in religious practice and volunteering. That this connection is less strong in Ireland may be accounted for by the embeddedness of churchgoing and religion in the lives of older adults. Respondents who go to Church a couple of times a week may be displaying a more purposive approach to how they choose to spend their time than either daily or weekly churchgoers. Perhaps it suggests an effect of being a person who, while very committed to Church, is measured in their religiosity, neither formulistic nor pious in their religious practice. While the value base of Christianity strongly urges practical social concern for others, undertaking voluntary work is a conscious action that involves making room for it among one's other concerns. It was interesting that no significant correlation was found between volunteering and the importance of religion or spirituality. This may reflect the different interpretations of what respondents understand by 'importance of religion' and the inadequacy of direct survey questions to elicit some of the subtler details of religious and moral frameworks.

Table 6.2: Church attendance by volunteering

	% of respondents		
	Voluntary Work		
Attendance at religious services	Yes	No	n
Daily	32	68	50
2–3 times per week	52.9	47.1	17
Weekly	30.2	69.8	63
Less frequent than weekly	21.1	78.9	19
Listen on radio/TV		100	10
n	48	111	159

Six missing values
Chi-square significance: p = 0.0

Christian values and volunteering

A strong sense of Christian duty or a disposition that was compatible with Christian concern for one's 'neighbour' was evident in the voluntary work of many older people in Rathmore and Rathbeg. For some volunteers their Christianity was explicit in the terms in which they spoke about their work. For others their Christianity or humanistic ethos was implicit in how attuned they were to the lives of others in their locality and their willingness to undertake practical work to help. One of the respondents in Rathmore who made a significant contribution in his locality by organising social events and by improving the environment through cultivating plants and flowers appeared motivated by a strong sense of Christian duty. He spoke about how he had coped with considerable personal difficulties and the degree to which he was motivated by a strong sense of God in his life. His work was described in religious/vocational terms by the religious Minister in the Parish as follows: 'God bless [name of person's] Ministry at [name of area].' His work was cited as an example of positive involvement in the community by other residents who made comments such as: 'There's one man in the complex and there's not enough hours in the day for all he does …'.

Another informant who was very community oriented came across as deeply spiritual and while she attended Mass she was critical of the institutional Church and frustrated by its unwillingness to allow women a fuller voice. A volunteer who did considerable work through his local Church and who was a regular Mass attender but not, he stressed, a daily Mass goer, said that spirituality and not religion was important to him.

One of the main organisers of an initiative to develop sheltered housing and day care facilities in the environs of Rathbeg spoke about his reasons for taking on the task:[9]

> I saw lovely older people who felt in the way in their own homes … Rejection is the worst thing that can happen to anyone … I wanted to make sure people felt wanted again and never feel a nuisance. They feel they are important now – instead of a nursing home they got a new life and made new friends.

On a visit to this centre we were impressed with the facility that has resulted from the vision, commitment and energy of this informant and other volunteers as well as the warmth and friendship we observed between this informant and the older people using the centre or residing there.

An organiser of an Active Retirement Association in Rathmore displayed practical social concern in her efforts to involve older people from the surrounding neighbourhood who had a quiet disposition and were not outgoing:

> I have to show I'm enthusiastic and draw them in. I tell people 'Right, come along, there's lots of different activities and you don't have to take part [actively].' Some people just like to feel part of the group, just to sit and listen and not feel lonely (Interview with Secretary of ARA, 17/9/03).

What stands out from all these examples of volunteering is that the respondents in question are attuned to the lives of other people. They perceive that there are people in their locality who are in some way excluded and who are not getting the best out of life and their motivation is to make the lives of those people better. They therefore

9 Interview and visit to day care centre 20/8/2003.

demonstrated a disposition that is compatible with Christian concern for one's 'neighbour'.

Two other factors that appeared to be important in leading people to become volunteers were, 1) personal circumstances which emphasised their own vulnerability, and 2) organisational ability. People spoke about how difficult personal circumstances had influenced their commitment to a particular form of voluntary work, for example the experience of widowhood, of being a single parent at an earlier stage in life, marriage break-up and in one case the experience of being wounded and looked after in a war situation. We observed that many of those in leadership positions in voluntary organisations appeared to possess considerable organisational skills. Some had skills from previous work, while others had developed their skills through becoming involved in their communities and in some cases availing of community education opportunities. This echoes an observation made by Abrams on the basis of his empirical studies of neighbouring that training or an acquired capacity for appropriate sorts of action is conducive to generalised reciprocity. The example Abrams gave was of people previously trained as nurses helping in Good Neighbour schemes (Bulmer, 1986: 114).

Volunteering and intrinsic satisfaction

We noted the extent of time commitment given to voluntary work, sometimes over a period of thirty years, in meals on wheels provision, Senior Citizens' clubs and the St Vincent de Paul society, and the considerable organisational skills and time commitment of people who served on committees that set up day centres in Rathbeg or worked on Care of the Aged committees or undertook local development projects. To what extent does volunteering enhance the volunteer's enjoyment of life? The experiences of volunteers are not always positive. As we discussed in Chapter Three we encountered considerable grievances among the volunteers in the meals service in Rathmore. The volunteers perceived that years of unpaid work had not been appreciated. A study suggested that the two most common drawbacks of volunteering as perceived by volunteers were that too much is expected of volunteers and being taken for granted (*Tipping the Balance*, 2002: 35).

Despite perceived drawbacks, the evidence is that volunteers derive considerable satisfaction from their work. When we examined the relationship between volunteering and enjoyment of life we found that respondents who did voluntary work were more likely to say that they enjoyed life very much (Table 6.3). However, the relationship, while strong, was not statistically significant. What did appear particularly striking was that only one of the 14 respondents who said they enjoyed life 'sometimes' did voluntary work.

Table 6.3: Involvement in voluntary work by enjoyment of life

	% of respondents			
Involved in voluntary work	Enjoy life very much	Enjoy life quite a lot	Enjoy life sometimes	n
Yes	64.6	33.3	2.1	48
No	48.3	40.5	11.2	116
n	87	63	14	164

One missing value
Chi-square significance: $p = 0.065$

A statistically significant relationship was found between volunteering and positive attitudes towards retirement. Respondents who did voluntary work were more likely to agree with the statement: 'Retirement has given me a new lease of life' (Q.31) and less likely to disagree (Table 6.4 below).

Table 6.4: Involvement in voluntary work by attitude towards retirement

	% of respondents				
	Retirement has given me a new lease of life				
Involved in voluntary work	Agree	Dis-agree	Don't know	N/A	n
Yes	66.7	16.7	8.3	8.3	48
No	42.2	31.9	20.7	5.2	116
n	82	45	28	10	164

Chi-square significance: $p = 0.014$

We found that being a member of a club or group was also significantly associated with positive attitudes to retirement. Volunteering and membership of clubs and groups may have been more strongly associated with positive attitudes to retirement than with enjoyment of life because such involvement provides a sense of having something worthwhile to do and consequently a meaningful social role in retirement. The question about enjoying life may have reflected a number of aspects of personal satisfaction and sense of well-being including satisfying personal relationships. In Chapter Eight we will combine a number of indicators of connectedness and examine how these relate to satisfaction with and enjoyment of life.

Informal helping

While only a minority of any age group undertake formal voluntary work, people help each other informally in many different ways. Informal help can involve practical instrumental tasks, giving emotional support or demonstrating solidarity in other ways. We firstly asked respondents about any help they gave to people outside their household (Q.14 a, b and c) and secondly, about help they themselves received with day-to-day living (Q.16). Approximately 40 per cent of the respondents (67 persons) said they helped someone outside their own household while a total of 14.5 per cent (24 respondents) said they received help with daily tasks from someone outside their household.

Table 6.5 below indicates that neighbours were reported to be the category that respondents helped most frequently (34 respondents) followed by help to sons or daughters (25 respondents). In eight cases help was given to other relatives while three reported giving help to friends. The level of help reported as given to sons and daughters may have underestimated help actually given such as childminding since not all respondents may have interpreted this as 'help or assistance with day to day activities to anybody not living in their household'.

Table 6.5: Who respondents helped outside the household

Help provided to	No. of responses
Neighbours	34
Daughters or sons	25
Other relatives	8
Friends	3
Parent	1
People in the community	1
Total no. of responses	72

n = 67 respondents who helped someone outside their own household
Five respondents said they gave help to two of the categories from the list

Respondents were asked about the nature of the helping tasks they did. They were free to mention more than one from a list of everyday tasks provided in the questionnaire (Table 6.6 below). Minding grandchildren and visiting were the most frequently mentioned helping activities. In most cases visiting involved visiting a neighbour to offer companionship, especially those who had reduced mobility or in some cases it involved visiting a neighbour in hospital. Tasks done for neighbours ranged from more instrumental help to more general 'watch and ward' activities. Instrumental help given to neighbours included doing household chores, doing errands such as collecting medical prescriptions, carrying household refuse bins in and out, giving lifts, looking after a pet, shopping, providing a meal, and doing maintenance jobs such as gardening or repair work. Many spoke about helping if someone was ill. Help related to personal care was less frequently mentioned than other tasks. Some people stressed that they were available to help if neighbours asked them. Comments included: 'I visit neighbours who have gone to a home or to hospital.' 'I bring in bins for the whole block.' 'We mind each other's houses.'

Table 6.6: Tasks that respondents did for others outside their household

Tasks done for others outside the household	% of respondents
Visiting	43.5
Childcare	41.9
Giving a lift / bringing places	14.5
Doing errands, e.g. library, post office, bank	14.5
Household chores	14.5
Looking after a pet	11.3
Other tasks	11.3
Shopping	9.7
Providing a meal	8.1
Personal care tasks	4.8
n	62

n = 62 respondents who helped someone outside their own household
Missing values = 5

The responses to Question 16 on whether anybody other than a home help came in regularly to give respondents help with daily activities show that the provider of assistance in most cases was a daughter (eight respondents) followed closely by a neighbour (seven). Help given by daughters will be examined in the context of kinship based reciprocities and interdependencies in Chapter Seven. The responses verified that help given by neighbours ranges from the provision of companionship to provision of services such as transport and may even include personal care task:

> 'Neighbours usually offer to bring me places.'
> 'A neighbour comes to do gardening and one to do painting.'
> 'A neighbour washes my toes, and cooks for me occasionally. She stitched my trousers. I get brown bread from another neighbour.'
> 'They [neighbours] call into me ... I can't go out. One phones me every night. They're very good to me.'

In the rural locality of Rathbeg a similar helping ethos exists. Tasks that neighbours do for each other are similar to those reported in Rathmore and people appear to value this neighbourliness. One respondent said : 'I have very, very good neighbours both sides ... people are more inclined to look out for each other in this area.'

Neighbours call to visit older people especially if they live alone and do errands or bring food if someone is sick. There is a particularly strong ethos of practical help and support given when someone is terminally ill and at funerals. Neighbours, including older people, sit up with someone who is dying to give family members a break. When there is a 'wake'[10] in a house neighbours frequently take on all the work of making tea for callers and preparing meals for the family. We found characteristics of positive neighbouring in both Rathmore and Rathbeg that were similar to those described by Bulmer ranging from the relatively minor, such as lifting bins to 'keeping an eye' on a neighbour to persistent caring and support. We will discuss positive neighbouring as a form of connectedness in Chapter Seven.

10 A custom in Ireland, particularly in rural areas, where a dead person remains in his/her home for one or two nights prior to burial and people come to the house to pay their respects to the deceased person and to offer condolences to the relatives.

Summary

In this chapter we identified forms of connectedness that gave meaning to the lives of older people in our two localities. In particular we examined volunteering and civic involvement through church, voluntary groups, community development activity and informal helping. We found as we expected that religious practice was strong and that a considerable amount of voluntary work and social action takes place through religious organisations and at a Parish level. We found that regular church attendance was very high among older people in both Rathmore and Rathbeg. Religion retains both social and personal significance for the majority of older people as it does for significant numbers of the general population in Ireland. Mass going and prayer for example appeared to give consistent meaning to their lives. Furthermore, for many people who went to Mass daily or a few times a week Mass was built into the structure of their day and had a social aspect to it. Older people are well represented in the composition of the many groups active in Parishes. However, we noted that church based groups were overwhelmingly dominated by white-collar respondents. We also discussed the likelihood that the high levels of church attendance in Ireland may not continue into the future.

The rate of volunteering among older people in Rathmore is somewhat higher than the average for this age group, with 30 per cent volunteering in at least one organisation. While it has been suggested that this represents something of an increase in volunteering among the older age group in recent years, it has also been suggested that the potential pool of volunteers among those who are retired is not being sufficiently tapped at present. This is interesting in the light of our findings that both volunteering and involvement in clubs or groups was significantly related to positive attitudes to retirement. While the voluntary work undertaken spanned the broad spectrum of society and included work in sports organisations, charitable groups and local community enterprise and development activity, much of it involved self-help associations for mutually enjoyable social and recreational activities such as Active Retirement Associations and more traditional

Senior Citizens' committees that provide meals, support services and organise entertainment for older people. We discussed the different ethos and structure of these two types of groups and we noted that some of the traditionally structured Senior Citizens' groups in Rathmore have not built in processes for continuity and renewal. We discussed a number of more recent community development approaches involving older people working together for example the project 'Voice of Older People'.

In addition to formal volunteering we found a strong informal helping ethos in all parts of Rathmore and Rathbeg involving tasks such as doing errands, bringing food, giving lifts, helping with small household chores or garden maintenance. Visiting among neighbours was practised, especially to people experiencing some degree of incapacity and living alone, and help was especially offered and appreciated when someone was ill. We came across many examples of supportive relationships, considerable amounts of practical help and strong civic commitment.

Chapter Seven
Relationships with Family, Friends and Neighbours

Introduction

Having suggested that involvement with others through purposive and communal leisure, informal helping, and voluntary and community activities represent important forms of connectedness, we now wished to examine the lived experiences of our respondents in terms of how connected or disconnected they appeared to be. This chapter examines the diverse patterns of relationships between groups of older people and their families, kin, friends and neighbours. It also explores the complex ways in which kinship and community can be experienced in the older years. We will explore the interdependencies and reciprocities among older adults in Rathmore and Rathbeg in the light of concepts and models related to kinship, community, social networks, neighbouring and friendship.

It would be misleading to present a picture of active, fulfilled and socially committed older adults as representing social reality for all our respondents. Levels of community participation were not uniformly high. Some parts of Rathmore appeared to have lower levels of social integration than other parts. Many of the social groups were predominantly female and friendships appeared particularly significant among women. While volunteering was higher in Rathmore than the national norm for this age group, it still involved a minority of older people. Furthermore, participation in collective undertakings in the community generally implies a certain level of good health, mobility, opportunity to develop significant personal relationships, opportunities for civic involvement and positive disposition towards communal activities in one's community. Among a heterogenous group of older adults one would expect to find different ways of relating to one's locality and to the broader civic community.

We wondered about older people who were not involved in the groups we identified in the previous chapters. Were there significant numbers of older people who were not connected with others in their locality through shared interests, civic involvement and friendships? If there were, what did this lack of social participation imply? Were they people who were self-sufficient, who led contented lives within their own family networks, who pursued their own interests and pastimes and did not wish for wider or more extensive contact? Or, were there groups of older people who were socially isolated, lonely and who lacked meaningful company and social engagement? Did such people belong to particular categories such as older men or were they more likely to be found in particular settings such as sheltered housing? Were older people who experienced health or mobility problems more likely to be less connected? What about older people who lived alone as single, widowed or separated persons – were they less connected within their communities than married people? Were there people who had experienced cumulative life disadvantages which in their older years had left them socially isolated and with little satisfying leisure or opportunities for social participation? In short, we wished to know what types of relationships and commitments were significant in the lives of different subgroups of our respondents.

The chapter begins with an examination of the concept of kinship. Changes in the nature of kinship ties are discussed and kinship as a basis of identity and support is compared with other social relationships in contemporary societies. The complex ways in which kinship, friendship and neighbouring can be experienced are considered as well as the impact of broad forces of social and economic change. The recent Irish literature on relationships among kin, neighbours and friends is discussed. In particular the links between such relationships and levels of social integration are examined. Data is presented on both types and levels of contact and support across all the relevant relationships. We describe helping, confiding and visiting interaction in Rathmore and Rathbeg. We highlight similarities and differences in social networks in the two areas.

The patterns of interdependencies and sociability reveal great variation in relationships with kin, friends and neighbours. We examine the experiences of older people who have extensive kin and community relationships as well as those who experience loneliness or social isolation. We also examine the circumstances of respondents who appear somewhat socially isolated and lonely. We briefly compare our findings with studies in other countries on the social networks of older people.

Investigating kin and non-kin relationships

All the literature had suggested that relationships with kin are one of the principal ways in which older people are socially connected. The evidence suggests that family remains the most important source of support for older people (Sundstrom, 1994, Grundy, 1995, Evason and Robinson, 1996). However, while the role of family is crucial and often the bedrock of support in old age, ties beyond the family are also important. It has been suggested that while families provide support they more rarely provide companionship. The role played by friends and neighbours in the support networks of older people has been highlighted in many recent studies. For example both Jerrome and Willmott have drawn attention to the importance of friendship and the value of clubs, religious groups and public places in the lives of older people (Jerrome, 1992; Willmott, 1987). Therefore, the evidence of the continued centrality of the family should not be taken to imply that we do not need to recognise that changes to the family as a social group have impacted on the nature of family support but that a broader understanding of social relations in old age should be taken. Researchers have identified a growing desire for autonomy and independence on the part of older people compared with previous generations, and have advocated that greater attention should be paid to different types of sociability and support.

Evidence suggests that interdependencies among older people are as complex as among any other age group. For example, Whelan and Whelan in their study of almost 500 retired people in Ireland have shown that the nature and quality of social networks which older people are involved in are as varied and complex as any other section of the population (Whelan and Whelan, 1988). The literature also suggests that there is a stronger sense in which family relationships have to be 'managed' than in the past, that even relationships with kin which may be assumed to be obligatory have to be negotiated and that there is now a more voluntaristic element in social relationships in old age (see for example Phillipson *et al.*, 2001 and Finch, 1989). The role that older people play in providing both emotional and practical support within their network has been emphasised. Recent research in Britain in three urban locations concluded that 'In most of the localities, and on most of the different types of support, older people retain ties of interdependence rather than dependence in their relationships within the family and beyond' (Phillipson *et al.*, 2001: 129).

It was not surprising, therefore, that we found great diversity in the forms of sociality and patterns of connectedness among older people in Rathmore and Rathbeg. We have already established that there was considerable variation in regard to leisure interests and communal involvement and our sample reflected great diversity in life circumstances. Our principal source of data on relationships with kin, friends and neighbours was the interview-based questionnaire survey. We also obtained accounts of such relationships from our interviews with key informants and we observed interactions in various social settings. We wished to get a general sense of which relationships were important, for what purposes and the nature of the reciprocities, support and friendships involved. We did not ask questions about the *quality* of relationships with other household members nor about the *emotional tenor* of their relationships in general. We were more interested in examining patterns of relationships that involved solidarity, shared interests and practical assistance. Nonetheless, we did get a sense of which relationships and activities contributed to overall satisfaction and meaning in life. In order to build up a picture of contact with relatives we established, through the interviews, where the respondents' close relatives lived – locally, elsewhere in Ireland or abroad. We asked about visits from and visits to significant others including sons or daughters, other relatives, friends and neighbours. We asked our respondents whether there was somebody in whom they could confide and whether anybody confided in them.

Kinship in modern societies

Much of the debate on the nature of modern solidarities revolves around the extent to which kinship ties have become attenuated. Janet Finch argues that the distinctive contrast in social relationships is between kin and non-kin (Finch, 1989: 234). In her explorations of the nature of kinship obligations in contemporary British society Finch concludes that most people do develop, with at least some of their kin, relationships that are characterised by the giving and receiving of support and which they understand to be qualitatively different from other relationships (Finch, 1989: 217). She refers to a need for an identifiable social place to which people feel they belong in an increasingly fluid and impersonal social context. She suggests that kinship does create such a place, although not necessarily exclusively so (Finch, 1989: 234).

Finch argues that a distinctive morality marks the boundaries between kin and other relationships. However, she suggests that it is a morality rooted in real ties between one person and another and the social meaning which these give to individuals' lives. This morality, which refers to a sense of obligation to give support, needs to be built up over time, reaffirmed and reinforced if it is to be sustained. Furthermore, she points out that the ways in which it gets translated into actions and commitments will be affected profoundly by the material conditions under which participants live. Demographic structures, the economic climate, the law and public policy all shape the need for kin support and the capacity to provide it (Finch, 1989: 236).

Membership of a kin network has one feature not shared by other groups to which people belong – you are a member and remain so whatever else happens in your life regardless of whether the bonds are close or distant. Irrevocable membership of a family of origin places each person in a series of two-way relationships with a number of individuals – mother, sister, aunt, cousin and so on. However, Finch points out that these are not conducted in isolation but develop as part of a set of relationships within the kin network as a whole and involve negotiations and reciprocities (Finch, 1989: 234–235). Furthermore, while she contends that kinship and friendship are distinct, she suggests that kinship ties do not necessarily imply active contact, that not all relatives are regarded in the same way for purposes of support and that factors such as gender will mediate the expectations of kin support (Finch, 1989: 233–234). Finch therefore sees the negotiated element of support among kin as a central element of family relationships. The concept of friendship has been proposed as a qualitatively different type of relationship to that of kinship and more attuned to conditions of choice, privacy and freedom associated with contemporary society. We examine the nature of friendship in Chapter Nine.

Relationships with kin in Rathmore

In Chapter Two we referred to evidence from Irish research that high levels of contact exist with kin across classes. In both Rathmore and Rathbeg there was evidence of local extended family where respondents had both children and siblings living in the locality. However, due to migration patterns the dispersed extended family was also evident and many respondents had children and siblings living in different parts of Dublin, elsewhere in Ireland and abroad. Emigration from Ireland was very high during earlier stages of the life course of our respondents. In contemporary Ireland involuntary emigration is uncommon although many young people in their twenties choose to live abroad for a period.

Since our focus was on interactions and help given and received outside one's own household, we did not examine the affective quality of relationships within households nor the reciprocities and sharing of help among household members including between spouses. However, we did ask about help given to someone within the household who had a specific dependency or disability. We found that 20 per cent of the sample gave help to a family member living with them, three of whom had done so until the person died quite recently. Spouses and sons or daughters were the most likely recipients of household help. In one case a sister was mentioned and in one case a parent. The nature of the dependency which necessitated help was either illness or mental/ physical handicap. The nature of the help given in these situations can be summed up by one respondent who said in relation to her son who had a disabling illness: 'I do everything.'

The marital status of our respondents in Rathmore was as follows: half the sample were married, 38 per cent were widowed, 9 per cent were single, and just under 3 per cent were separated. A total of 37 per cent of the sample lived alone while 63 percent lived with at least one other person. All the Rathmore respondents were born in Ireland, however, only one tenth of the sample were born in the same general area of Dublin. The majority were born in another part of Dublin or elsewhere in Ireland. Several had worked outside Ireland and a considerable number had sons or daughters living abroad. Most of the sample had large numbers of lateral and/or descent kin. There was just one respondent who had no relative. Respondents were asked if they had any close relatives living within two miles or thereabouts. Close relatives were defined as brother, sister, grandchild, aunt, uncle, nephew and niece. Almost two thirds had a close relative in the locality while over one third did not. The majority of respondents who had close relatives in the locality had between one and three relatives.

Contact with and support from and to children

We elicited data on respondents' relationships with their children and on the exchanges between them from a number of questions including frequency of contact, both face-to-face (Q.25a and b) and telephone (Q.27a), help received with daily tasks (Q.19) as well as help given (Q.14), events and occasions regularly enjoyed and anticipated (Q.34 b) and who their confidants were (Q.20). We could not access on a large scale the details of the affective quality of relationships between parents and children. Nonetheless, some respondents did talk spontaneously about their children and we noted these comments.

We asked our respondents about visits from and to significant others. Table 7.1 summarises the frequency of visits from significant others, showing only weekly visits and no visits. Visits from sons and daughters were the most common visits our respondents received. Furthermore, children were the most common people who confided in the respondent and in whom they confided. Daughters in particular were an important source of practical regular help and respondents helped their children especially by minding grandchildren. Family occasions, especially those involving grandchildren, were frequently referred to.

Table 7.1: Summary table of frequency of visits from significant others

	Frequency of visits %	
Visits from	Weekly	Rarely/never
Children	85	0.7
Relatives	22.4	32.8
Friends	43.6	29.7
Neighbours	46.7	36.9

n=165

Almost 85 per cent of those who had children said that their children visited them at least once a week and indeed many said that one or more sons or daughters called every day. Children visited their parents more frequently than parents visited their children. A total of 58 per cent said they visited their children's homes at least once every week and 23 per cent visited at least once every month. Approximately 12 per cent visit their children infrequently (a few times a year). Those who visit their children infrequently were much more likely to have a major illness or disability – seven out of nine who said they visit their children infrequently had a major illness or disability. In some cases the children were living abroad, making regular visiting impractical. Telephone contact was maintained with children living abroad and several respondents spoke about a forthcoming planned visit to a son or daughter. (Question 34c). In a very small number of cases marital conflict may have damaged relationships, however, we did not ask about these aspects of people's lives.

Almost one quarter (24 per cent) of the sample had unmarried children living with them and we noted that a few of our respondents were continuing in the role of full time homemakers for their adult children still living at home. Even when children had left home we noted that the domestic role continued to be a major source of identity for women revolving around cooking and caring. This can be understood in terms of reciprocities and commitments built up over time which have created an enclave of sociality within which expectations about giving and receiving help become established.

Children were also major confidants in relation to having someone to discuss problems with (Q.20a). When we specifically asked our respondents who, if anyone, they confided in if they had problems or worries they were more likely to say a daughter, a spouse, a son, another relative and a friend in descending order. The most common person with whom a respondent discussed personal problems was a daughter; 71 respondents mentioned a daughter or daughters representing 57 per cent of respondents who had daughters. Sons were mentioned by 49; this represents 41 per cent of respondents who had sons. Spouses were mentioned by 54 respondents. Underlining an important theme throughout the study, that of the contribution of older people themselves to the lives of others, we found that 72 per cent of respondents said there was someone who relied on them for advice or support (Q.20b).

Several respondents said that their children were 'very good.' We have already noted that daughters, in particular, were relied on for regular practical help: 'Two daughters come over once a week, they couldn't do enough for me.' 'It was suggested after I had my stroke that I might need a Home Help, but I didn't need one, my daughter helps.'

On a number of occasions when we were interviewing we observed daughters calling to bring their mothers shopping. This finding accords with studies in other countries which have found that spouses and daughters in particular are the mainstay of practical help in old age (See for example Phillipson *et al.*, 2001; Wellman, 1990).

When we asked our respondents about future events that they were looking forward to (Q.34c), many of them spoke about family occasions such as family dinners, and special occasions involving grandchildren such as communions, confirmations, graduations and birthday parties. One woman said: 'Yes, two '21sts', grandchild's graduation, there's always a confirmation or communion. It's lovely ... kids' concerts and all the rest ... it's nice ...'.

A few were looking forward to visiting sons or daughters who lived abroad, such as planning a trip to Australia, where a lot of young Irish adults are living. Of our respondents who had grandchildren one quarter had grandchildren living abroad. One respondent spoke about how much she enjoyed her grandchildren and described activities such as playing pitch-and-put and going swimming with them. But, she added, 'it was a feast or a famine' situation in terms of spending time with them as they lived some distance away. Another respondent said he enjoyed bringing his grandchildren to school every day. The interviewers in the thumbnail sketch that they completed commented on how 'family oriented' many of the respondents were.

The telephone is a major point of contact for older people. Many respondents spoke about being in daily telephone contact with their children. For people who were somewhat incapacitated the phone is clearly a major asset. Women tend to use the phone more than men for general sociability (Fahey and Murray, 1994). One male respondent said: 'No one rings me. I leave it to the wife ... when I'm home I don't touch the phone ... the wife phones friends.' Another said 'When the [adult] children call they always look for mother.' A number of the respondents told us they also kept in touch with their children by e-mail.

However, not all contact with adult children is necessarily perceived by older people (or the adult children) as positive. While adult children have been identified in the literature as an important part of the emotional support system of older adults, the evidence suggests that contact has both positive and negative aspects. Two Irish studies have suggested no positive relationship between morale among older people and interaction with children (Fahey and Murray, 1994: 125, Whelan and Whelan, 1988: 60). Whelan and Whelan suggest that this can be interpreted in terms of the strong elements of obligation associated with kin relationships and the resolution of power/exchange imbalances involving critical resources of health and income. A theoretical possibility advanced by Whelan and Whelan (which they were unable to confirm within their study) was that where parents were able to choose the conditions under which interaction took place, illustrated by for example situations where they visited children rather than children visiting them, contact would be expected to have positive effects. We were also unable to verify such a hypothesis given the focus of our study. However, one remark made by one of our informants suggested that she valued being a certain physical distance from her adult children. This informant when talking about her children and where they lived in Dublin added quietly that these locations were 'near but not too near' which she implied suited her.

Grandparenthood

While the experience of grandparenting was not an area that we specifically pursued in the study, we obtained insights into the experiences of older people who were grandparents. Recent research in Ireland and Britain on the meaning of grandparenthood and its contribution to older people's lives found that grandchildren were important in the lives of grandparents (Lundstrom, 2001, Clarke and Cairns, 2001). A survey of a national sample of grandparents in Britain found that the majority of grandparents (62 per cent) saw grandchildren at least once a week and while weekly contact was related to proximity other forms of contact were reported by a similar percentage through letter, telephone and e-mail. Qualitative interviews with 45 grandparents confirmed the importance of grandchildren to grandparents as expressed by emotional closeness and contribution to the quality of their lives (Clarke and Cairns, 2001). The authors expressed themselves surprised by how much grandfathers were involved in the lives of their grandchildren. The study highlighted variation between grandparents in what they provided for grandchildren, how much they helped with grandchildren and what they were willing to do for their families. While grandparents were generally prepared to step in to help their children with grandchildren when needed, they also expressed the wish to maintain their own lives and interests. However, the authors also discussed the extent to which grandparenting is a role characterised by negotiation and constraint. 'Negotiating' and 'not interfering' were key themes in grandparents' discourse.

At the time of this study the only Irish study on grandparenting is a qualitative study based on interviews with 58 grandparents and while the study does not purport to be representative, it does propose a classification of grandparents based on their degree of involvement with their grandchildren (Lundstrom, 2001). The study also suggests what trends may be emerging in relation to intergenerational family bonds in Ireland. Most grandparents in the study lived within 10 miles of their grandchildren and, on average, saw most of them at least once a week. One third of the grandparents had some grandchildren who lived abroad (Lundstrom, 2001: 36). The author found that there was close geographical proximity between many grandparents and grandchildren and that this led to frequent contact which gave great joy to most grandparents. She noted that geographical proximity did not appear to be typical of other countries, however, there are indications that geographical proximity will decline in the future due to housing and traffic factors making it more difficult to maintain regular face-to-face contact (Lundstrom, 2001: 52).

Another theme that emerged in Lundstrom's study was a shift in the balance of power among generations. Many grandparents made a conscious decision not to interfere in the lives of their children, particularly in relation to discipline and child rearing (Lundstrom, 2001: 33). The author suggested that this appeared to represent a change from the elder-dominated structure of the Irish family at an earlier historical period (Lundstrom, 2001: 52). Furthermore some grandparents believed that a further shift in the balance of power had occurred from parent to child. Specific concerns that grandparents had related to drug and alcohol abuse in particular, and to a lesser extent materialism, too little discipline, sexual promiscuity and the detrimental effects of television (Lundstrom, 2001: 39).

The themes identified in these studies resonate with concepts in the literature that have already been discussed such as that of 'intimacy at a distance' and 'involved disinvolvement'. We have noted comments about how much respondents enjoyed activities with and special occasions involving grandchildren. At the same time many spoke about allowing adult children to live independent lives and the importance of not interfering in their children's lives: 'We don't live in their pockets, we never interfere ... we don't want to torment them.' Another respondent spoke about her children being a 'fantastic support' at a time when she was ill, however she added: 'but they lead their own lives'. Another respondent, a widow, talked about feeling lonely sometimes but having to maintain some distance from her married children: 'after rearing a family you'd miss them, [but] you can't be ringing them all the time'.

An insight into the burdens of grandparenting was obtained through interviews with key informants. We learned from voluntary and community workers in Rathmore that grandparents, principally grandmothers, were very involved in helping to rear grandchildren. In a more working class part of Rathmore this was said to represent a considerable emotional and psychological burden since many of the grandmothers involved had already reared large families, often in difficult financial circumstances and in a culture where heavy drinking was the norm, especially among men.[1] One voluntary worker suggested that grandparents were now taking on problems of their adult children and grandchildren associated with marriage separation, excessive drinking, drug addiction and crime. While we did not obtain any direct evidence on the stresses of grandparenting, we found evidence of less connectedness in some parts of Rathmore that may be associated with lower levels of social capital in less socially advantaged localities.

Relationships with kin outside the household

Regular visits from other relatives were much less frequent than visits from children, between one fifth and one quarter had a weekly visit from a relative. This was considerably less than the 46.7 per cent who had a weekly visit from a neighbour and the 43.6 per cent who said they had a weekly visit from friends (Table 7.1). There was great

1 Interview with voluntary worker 19/11/02.

variation in frequency of contact with relatives. While over 40 per cent had a visit from relatives at least monthly, approximately one quarter received a visit a few times a year, 16 per cent said they rarely had a visit while 17 per cent said it did not apply, either because they had no relatives or no relatives with whom they had contact. Visiting relatives was usually reciprocal. While it was difficult to tease out the meaning of both frequent and infrequent contact and visiting, we noted some characteristics of the group who were not on visiting terms with relatives. In some cases the relatives with whom they had contact had died. On the other hand it was clear that visiting contact was not automatic even when relatives lived within the locality. As far as we could ascertain non visiting was not significantly related to poor health. However, a higher proportion of those who were widowed were less likely to have visiting contact with relatives than either married or single respondents. This suggests that widowhood may be associated with a loosening of ties with other kin. A higher proportion of those who were not on visiting terms with their relatives experienced loneliness sometimes or frequently.

The degree of continuing and regular contact, if any, depended on the number of relatives whom they had contact with who were still alive, where relatives lived and the degree of friendship involved. We interviewed people who were the last surviving member of a big family and people who still had large numbers of siblings living. One lady, a widow with no children and the only surviving person left from a large family said: 'I miss my brothers and sisters, we were a close family.' In contrast, another respondent had 14 brothers and sisters still alive. He described his contact with some of them:

> I have three brothers in the United States … I went to see them … the wife wouldn't come … I have a sister in (near Dublin) who I see once in a while. My brother was my mainstay but he moved to (elsewhere in Ireland) … at least I can visit him there … about twice a year … I visited him a few times. I stay away from the brother in (elsewhere in Ireland) who drinks too much. I went to visit my sister who lives in (place in Dublin).

Another male respondent spoke about his relationships with his brothers and the ritualised occasions when they would meet: 'I wouldn't talk to my brothers on a regular basis but we'd meet at funerals and so on and get on well. We don't socialise together but there are no problems ... if we needed each other we would help.'

Respondents often referred to a particular relative to whom they were close. One lady visits her sister who lives nearby every day. Another lady said: 'I don't visit my brother, I used to see my sister more but she moved to (place outside Dublin) so it's not as often now but we keep in touch ... She'd come if I needed her.'

A male respondent spoke about how involved he was with his wife's family whom he described as very close-knit. A number of single people said that they had visits from or made visits to nephews or nieces. One lady spoke about the good rapport she had with her niece, nephew and grandnieces: 'They like to hear stories about the past, but they might not want to hear all that went on!' While comments made about relatives were generally positive, there were also some negative comments. For example one single lady spoke about her nieces who she claimed did not keep in touch with her: 'I have nieces in the locality but they don't bother with me.'

One informant spoke about how relationships with siblings change and become less salient when people marry but are often restored or renewed when a spouse dies or if people separate.[2] This informant visited and gave emotional support to a brother in a nursing home who had poor health and was separated from his spouse. He spoke about the obligation to help your siblings: 'You wouldn't let down your own.' A number of respondents also spoke about receiving or giving support to sisters/brothers-in-law. Another interesting phenomenon we came across was that of older emigrants returning home to their local place to spend their remaining years with their siblings even though their own children and grandchildren remained in England or Scotland where they were reared.[3]

The data obtained on contact and support between kin resonates with much of Finch's research on kin support. The data reported here suggests that both contact and support are variable. Since reciprocities and obligations are built up over a long period of time and involve real

2 Conversation on 3/3/04.

3 Data from DIT student who interviewed older people in St Brendan's sheltered housing complex in Mulranny, Co Mayo in March 04.

ties between people it was not possible given the focus of this study to learn more about the nature of such reciprocities in the context of individual biographies. However, we noted a sense of obligation towards kin and an expectation that help would be forthcoming from kin if needed and indeed such help was often provided. At the same time, keeping an appropriate amount of independence from other adults is a guiding principle so that the desirable situation for many people is to never have to use the safety net of kin support (Finch, 1989: 240–241).

Relationships with friends

Friendship emerged as a central theme in our study and was strongly evident in the social interactions in social, leisure and community groups. We will examine the different contexts and different types of friendship in more detail in Chapter Nine. In this section we focus on data obtained in the survey, in particular reported visiting patterns involving friends. In doing so we must bear in mind that visits from and visits to friends are just one indicator of the significance of friends in the lives of respondents. Visiting among friends represents just one form of sociability and is often class related. Furthermore, friendships were reported with relatives and neighbours. Indeed some respondents made no distinction between neighbours and friends. Therefore, our categorisation of non-kin significant others into separate categories for 'neighbours' and 'friends' in the questionnaire must be regarded as somewhat problematic. In many cases the meaning attached to these categories of 'friend', 'neighbour' or indeed 'relative' was overlapping and probably more revealing of the significance of these relationships than the statistical analysis of number of visits. With these considerations and limitations in mind we will examine the survey data on contact with friends.

Visits from friends and visits to friends occurred more frequently than visits from/to relatives. Visits to or from friends occurred weekly for over 40 per cent of the sample and at least monthly for over 10 per cent of the sample. About 30 per cent did not have visits from friends while about 35 per cent did not visit friends. We found that in total about 41 respondents or one quarter of the sample did not visit or receive a visit from a friend. While visits involving friends occurred considerably more frequently than visits involving other relatives, slightly more respondents said that they would confide in relatives than said they would confide in friends (26 and 23 respondents respectively). It is worth noting that neighbours were not typically approached with personal problems, only 7 put neighbours in this category. Others mentioned more infrequently were doctor or nurse.

Some of the reasons given for lack of visiting involving friends included disability, lack of space to receive visitors (a particular difficulty for residents of sheltered housing) or meeting friends elsewhere: 'My friends are not able to visit. They're my own age in their 80s.' 'This place [sheltered housing] doesn't facilitate visiting … people need their privacy.' 'I have one very good friend. We keep in touch by phone … we meet in town.'

Overall, there did not appear to be a particularly strong emphasis on visiting friends at home or having friends visit one's home. In contrast, we found forms of friendship that essentially involved sociability and that appeared to be an important part of the enjoyment that respondents got from social interaction in clubs and shared leisure activities. Indeed we found that accounts of friendship in our study did not appear to correspond with a view of friendship as intimacy found in some current social theories – a theme we will examine further in Chapter Nine.

We found that the higher socio-economic groups were more likely than manual groups to be involved in visiting friends. Table 7.2 shows that receiving visits from friends was significantly related to belonging to a white-collar class.[4]

4 A similar significant relationship was found between social class and visits made to friends.

Table 7.2: Visits from friends by respondent's social class

	% of respondents		
	Respondent's social class		
Visits from friends	White collar	Manual	n
Weekly	49.5	35.3	72
Monthly	18.6	7.4	23
A few times per year	11.3	14.7	21
Rarely/NA	20.6	42.6	49
n	97	68	165

Chi-square significance: $p = 0.006$

The literature suggests that the middle class pattern of friendship formation is associated with the home being used for entertaining while working class sociability is more likely to be defined by its initial setting for example the pub or club (Allan, 1996: 87). We discuss the relationship between social class and forms of sociality in Chapter Nine.

While females were twice as likely as males to report having a weekly visit from a friend, a similar number of women and men reported not having a visit from a friend. The group who were not involved in visits with friends were somewhat more likely to experience loneliness than others although the association was not strong. However, there was a stronger association between visiting friends and health – those who were not involved in visits with friends were considerably more likely to report that their health was fair or bad than those who were involved in visiting.

Marital status was also related to visiting friends – we noted that only one single person was included in the subgroup of 41 respondents who did not make or did not receive visits from friends (there were15 single people in the total sample). On the other hand those who were widowed were disproportionately represented among the group not involved in visits with friends. This supports findings from previous Irish surveys (Layte *et al.*, 1999 and Fahey and Murray,

1994) which suggested that single elderly people may be more pro-active in making contact with others to compensate for their position.

Having friends outside one's immediate locality may be particularly important for single people as is evident in the following excerpt from an interview with a single lady who was described by the interviewer as 'very vivacious and with a great network of friends'. She described how she felt excluded by the married women in her area: 'They go around in clumps, in two's and three's ... those who came new and brought up families together ... , being an old maid I was the odd one out'. Therefore the meaning of community for this respondent was not limited to engaging with people in her immediate locality. Moreover, being a retired professional she fell into the category for whom according to Fischer 'education by itself meant broader, deeper, and richer networks' (Fischer, 1982: 251).

Relationships with neighbours in Rathmore

We saw evidence of a strong ethos of help among neighbours in Rathmore and Rathbeg in the many practical ways in which neighbours helped each other and the considerable value attached to good neighbours. Neighbourliness was demonstrated through friendliness, help with daily activities and support in times of difficulty. In many cases trusting and friendly relationships had developed over many years of living close together and going through shared experiences of rearing children, being involved in community activities, worshipping in church or having similar interests. We have already noted that of the sizeable proportion of respondents who helped someone outside their own household (67 respondents or 40 per cent of the sample) neighbours were reported to be the category that respondents helped most frequently.

Frequent visiting among neighbours coexisted with no visiting, however neighbours not visiting each other did not indicate any lack of contact. Approximately 47 per cent reported having on average a weekly visit from a neighbour and 43 per cent said they visited neighbours weekly. About one third of the sample (56 respondents) did not receive visits from neighbours nor did they visit neighbours. Many respondents were anxious to point out that although they did not visit neighbours they had regular contact with each other. Regular contact involved meeting on the street or over the garden wall or neighbours calling to each other for a particular purpose. Contact might also involve keeping an eye on a neighbour's house if they were away, holding a key or calling to inquire about someone who was ill: 'I don't visit neighbours but I see my next-door neighbour over the garden wall ... gardening is a common interest.'

An interesting insight into visiting or not visiting the homes of neighbours was given during a discussion between the author and the members of a Ladies' club in Rathmore.[5] One woman said that when she became widowed she did not like to visit a neighbour because she felt it would be intrusive with her husband there. However, when her neighbour also became widowed they began to visit each other regularly. Another member explained that because the club was small they were all friends but that a few would be closer to each other. Two or three spoke about an arrangement where they visited one another's house in turn once a week and said this had been going on for years. They agreed that not visiting neighbours did not mean not having contact but said that they would call to a neighbour's house for a purpose, for example to pass on news that they knew their neighbour would wish to hear.

Where visiting occurred it was seen as a positive thing and was a source of support. In some instances people's homes were very open to neighbours, in other cases people's home appeared very private. There were a small number of respondents who commented negatively on their neighbours and a small number who said there was not much neighbourliness in their area. One respondent said:

---

5 Presentation of findings and discussion with members of a Ladies' club in Rathmore on 24/11/04.

> There's no such thing as neighbours visiting here. I'm great with one neighbour but not so nice on the other side. You'd be dead and buried before anyone would know, people keep to themselves … , that's urban life.

In contrast another respondent said: 'I won't be dead a day before they find me.'

Several respondents spoke about neighbours who had died or moved out. When referring to changes in their area they often spoke about the 'old crowd' who knew each other and 'a younger crowd' who 'aren't very friendly and we don't know their names.' One respondent described how his road had changed: 'The place has changed – 50 per cent are new and I would know them to see … Once I knew 100 per cent of people here. We used to run concerts, it was friendlier then.'

However, there were several respondents whose experience was that their area was friendlier now that it was in the past. One widow explained her experiences:

> If anything it's more friendly now. When I was a young widow, the first in the area, women had their husbands and didn't understand. As there are more widows now it has got more friendly … , they understand now … , they're in the same situation.

Some attributed lack of visiting to their stage of the life cycle suggesting that they didn't see neighbours as often as when their children were small: 'Since I'm no longer involved in any communities there's no reason for people to be calling. Times have changed and we live our own lives … children have grown.'

The work of Abrams and Bulmer on the components of positive neighbouring suggested that on the one hand friendliness and helpfulness were important and on the other hand distance. This could be seen in the ethos of consideration and practical support among neighbours coupled with respect for privacy in one's own home. At the same time many neighbours who had been involuntarily placed together had developed close friendship ties over many years. The considerable amount of informal help and informal sociability that occurred among neighbours strongly suggests that visiting alone should not be taken as the only or indeed most significant indicator of neighbourliness. A desire for privacy and control over the circumstances of interaction with neighbours can coexist with reciprocities and altruism among neighbours.

Relationships with relatives, neighbours and friends in Rathbeg

Turning to our more rural community of Rathbeg we found that respondents who had children living in or near Rathbeg were visited frequently by them, at least on a weekly basis and sometimes on a daily basis. For respondents who needed assistance with aspects of daily living their children provided regular help with cooking, showering, shopping, bringing to Mass and doing errands. Children who were living abroad visited perhaps once or twice a year and many older people visit their children in different parts of Ireland and abroad. Respondents spoke favourably about the help their children gave to them.

Contact was maintained with relatives and face-to-face contact was more frequent if relatives lived in the locality. It was noted that cousins visited each other and were often regarded as friends and confidants. Cousins were especially important for single people; if their siblings were dead or they had no siblings their relationship with cousins was an important source of support. These relationships have been built up over time and involve reciprocities. One respondent said in response to Question20a about having someone to confide in: 'I suppose you would contact your own ones', referring to her cousins. Another respondent spoke about how she regarded it as her obligation to help to nurse a cousin who was terminally ill:

> I took it as my duty, I suppose maybe because she didn't have anyone else. I was full-time able to do it and I did it. I looked after other relatives who were ill, nursed them through illness and sat with them.

It was common for neighbours to visit each other and older people living alone or who had some degree of incapacity were visited regularly. We referred previously to a strong ethos of practical help and support offered when someone is terminally ill and to families who have been bereaved. Regular practical help is offered similar to Rathmore including giving lifts to Mass or bringing a person to the shops. We pointed out previously that official home helps are highly regarded by their clients and offer friendship as well as practical help.

In contrast, one or two respondents said that there was little visiting among neighbours in their immediate area. One respondent said that: 'They don't visit like they used to' and pointed out that most of her neighbours were out at work during the day. She also suggested that the practice of neighbours visiting varied in different parts of Rathbeg. We also noted that the St Vincent de Paul Society and the Legion of Mary visit people who live alone and who may be socially isolated as well as older people who are long term residents in the community hospital. However, it is worth pointing out a distinction that a senior local authority official made between a type of perfunctory 'visiting' that she claimed some voluntary groups do and what she called 'befriending'. At a feedback meeting in Rathmore the official suggested that some charitable groups who visited older people were just 'doing the rounds' but that what people needed was friendship.

Rathbeg has not been unaffected by broad processes of social change and this can impact differently on older people in different situations. Many travel regularly to visit their children and grandchildren in other parts of Ireland and abroad. On the other hand people with more limited kinship networks including single people who have lived locally all their lives are less likely to travel and may find that modern forms of leisure do not interest or involve them. One respondent, a widower, recalled fondly the practice of 'raking' or 'rambling' in his young days in Rathbeg. This referred to a common practice of visiting other houses in the area where someone would play a musical instrument and there would be dancing. Another single female respondent missed social activities such as card playing in a local hall that were customary in the past. While almost all the respondents use the phone regularly to keep in contact with people, not having a car limits one's social activities particularly in Rathbeg.

Loneliness among respondents

Loneliness is distinct from social isolation, which is understood as the lack of enduring ties with others. Loneliness is a subjective state, involving the way in which people perceive and experience their circumstances. Conceptualisation of loneliness sees it as having both a social and an emotional dimension (Weiss, 1973). Victor *et al.* identified three distinct definitional categories of loneliness: functional, a state of mind and related to friends network (Victor, *et al.*, 2002). Functional loneliness was defined in terms of a loss of a range of abilities, for example due to poor health. Loneliness as a state of mind refers to one's ability to fill time, to be contented when alone and be able to motivate oneself. In relation to one's social network, loneliness refers to closeness with others and having a confidant. A recent Irish study that sought to determine the prevalence and experience of loneliness and to identify preventive strategies found that the majority of older people in Ireland are not socially isolated and are part of a locally integrated support network (Treacy *et al.*, 2004). While levels of social and family loneliness were low among older Irish people, the authors found that romantic/emotional loneliness affects over half of older people in Ireland (Treacy *et al.*, 2004: 175). Higher levels of social, family and romantic loneliness were found to be influenced by many factors, including increasing age, being single or widowed, being educated to primary level only, in a lower social class classification, living in either rented accommodation or with relatives in a rural area, having perceived poor health and a lack of transport (Treacy *et al.*, 2004: 182). Fahey and Murray found that being widowed had by far the strongest association with feelings of increased loneliness followed by poor health as measured subjectively (Fahey and Murray, 1994: 121).

The research suggests that loneliness is an extremely complex phenomenon that can cover many different aspects including feelings of dissatisfaction, lack of fulfilment or social isolation. It may also indicate depression. While our interest in examining different forms of connectedness among older people implies a much broader approach to relationships and commitments than is generally undertaken in studies of loneliness and social isolation, we did include questions on loneliness and social isolation as well as on social contacts generally.

We asked respondents whether a week would go by without them visiting someone or having a visitor (Q.26). We found that twenty one respondents, or about one eighth of the sample said that a week would go by without a visit.[6] The respondents who appeared somewhat socially isolated included widowed, married and single respondents, people who lived alone and people who co-resided and women as well as men. Respondents were also asked if they experienced loneliness (Q.33). Two thirds said they rarely or never experienced loneliness (67.9 per cent). Only 5.5 percent said they were frequently lonely while 26.7 per cent said they were sometimes lonely.

We obtained insight into aspects of the respondents' experience of loneliness such as the restrictions chronic illness placed on daily activities, lack of meaningful participation and loss of a close relationship. What was particularly striking and echoed Fahey and Murray's study was that seven of the nine respondents in Rathmore who said they experienced loneliness frequently were widowed. Many spoke about being lonely following the death of a husband or wife, or of a friend or relative and one person who said she was frequently lonely said she wished she had her husband back. One person spoke about loneliness in terms of not being able to get out due to incapacity: 'I'd love to be able to get out but you know you're not physically able to'. One man who was in his 80s and was fit and mobile missed his past active involvement with many church groups: 'I miss the company of that active life … not being able to give as much'.

Some spoke about the strategy they adopted to deal with feelings of loneliness such as finding something to do, going for a walk, keeping in touch with people by telephone, turning on the television or going to the pub. Others spoke about activities that they did to prevent loneliness such as bridge and going on holidays while one lady spoke about how since she took up teaching art she never felt lonely. People

---

6 There was some degree of ambiguity in this question as a few people appeared to interpret it as any person calling while most took it to refer to anybody other than close family members.

spoke about particular times that they were more likely to experience loneliness such as winter, night-time or at weekends when children are with their families.

Patterns of social contact and social support in Rathmore and Rathbeg

The overall picture presented is of dense networks of interaction involving family, relatives, friends and neighbours in which older people are active donors as well as recipients. While some people had very small kin networks, few friends and infrequent social interactions with neighbours, only a very small proportion said they experienced loneliness often. The meaning of different types of social interactions with significant others is complex and we found tremendous variation in visiting patterns (Table 7.1). Yet, there were also some interesting patterns that give clues to different types of sociality.

A majority of those not in visiting contact with relatives were also not in visiting contact with friends. For example approximately 65 per cent of those who did not have visiting contact with relatives did not receive visits from friends. This compared with approximately 30 per cent of the general sample who did not receive visits from friends (Table 7.1). However, visits from neighbours occurred almost as frequently irrespective of whether respondents were in visiting contact with relatives or not. For example almost 43 per cent of those not in visiting contact with relatives received a weekly visit from neighbours compared with almost 47 per cent of the total sample.

This suggests that connectedness with relatives and friends may represent a different type of sociability than connectedness with neighbours. It seems possible that connectedness with neighbours is open to people irrespective of their cumulative life histories which may have left many people without significant relationships with friends or relatives. We found that relationships with neighbours could be of immense importance for those who had few relatives and friends. However, there were many people who had extensive kin, friendship *and* neighbour networks in our study areas. We will tease this out later when we go on to develop a classification of people based on different forms of connectedness.

In order to explore further the question of non-visiting contact and general experiences of life we examined those respondents who had no visiting contact with significant others including relatives (other than children), friends and neighbours. We found that 14 of our respondents (8.5 per cent of the sample) did not have any relationships of these kinds. None of the 14 was single, half were widowed and half were married. Almost two thirds described their health as fair or bad compared with approximately one third of the general sample. We found that they tended to experience loneliness more often, to be less likely to participate in regular enjoyable activities and to be less likely to say they enjoy life very much. Approximately two thirds said they experienced loneliness frequently or sometimes compared with approximately one third of the general sample. Only 7 per cent said they enjoyed life very much compared with over half of the general sample. Over 40 per cent said there were no events they did regularly for enjoyment compared with 27 per cent of the general sample.

Among this group who had little contact with people outside their immediate family, over three quarters said they had a weekly visit from a son or daughter while the remaining one quarter had a monthly visit. This suggests that regular visits from children is not a bulwark against loneliness and that companionship and sense of involvement may derive more from contact with peers including other kin, friends and neighbours. Where friendships have developed based on reciprocities and relationships of mutual respect the older person may be more active in choosing the terms of the relationship.

Not being involved in visiting is associated with lower levels of morale and satisfaction and may derive from a combination of factors including poor health, bereavement, life circumstances and personality characteristics. However, meaningful contact also takes place in communal places such as social centres, clubs, pubs and churches. We have demonstrated how such settings create enclaves of sociality within which friendships develop. The possibility of facilitating opportunities for peer friendships is an important challenge for public policy and we will return to this later.

Differences in patterns of kinship and neighbouring between Rathmore and Rathbeg

Were there differences in the social networks of older people in Rathmore and Rathbeg? Our study was not designed to make a detailed quantitative comparison of the social networks of older people in both localities. However, a number of key points of similarity and difference emerged:

1. There was a stronger sense of social/cultural change in Rathbeg than in Rathmore. This was reflected in respondents' spontaneous references to the changing composition of the community, changing social practices and leisure activities. In Rathmore while respondents made reference to changes in their neighbourhood, this related more to life course changes than to socio-cultural changes per se.
2. Social networks were large, heterogeneous and dense in both localities. The vast majority of older people were involved in dense networks of reciprocal relationships with family, kin, neighbours and friends.
3. Support from children was a strong feature of the lives of older people in both Rathmore and Rathbeg and was evident in the frequency of contact, help with aspects of daily living and enjoyment of family celebrations. Sibling support was important and especially for single persons who did not have children available to them. Support and friendship involving cousins and other kin was very evident in Rathbeg.
4. Neighbours were often regarded as friends in both Rathmore and Rathbeg. There were strong patterns of informal helping among neighbours in both localities. Help given by neighbours to people experiencing illness and bereavement may be a stronger feature of Rathbeg than Rathmore.

5. There are localities within Rathmore where cultural practices associated with traditional close-knit communities were observed.
6. There was a strong sense of a diaspora in Rathbeg. Many people returned to Rathbeg having lived abroad for some years or to retire, and many returned regularly to visit family. Communal events provide a focus for residents, people with family connections in Rathbeg and regular visitors.
7. There was a greater number and greater diversity of social groups in Rathmore than Rathbeg. However, there were more active groups catering for the interests of older people in some parts of Rathmore than in other parts. The lack of a satisfactory public transport service in Rathbeg limits opportunities for older people without private transport to develop and pursue interests outside their locality.

Comparison with patterns of kinship and friendship found in other studies

We will refer briefly to a number of studies from other countries that examined social networks. Phillipson concludes that studies of the personal networks of older people confirm that kin provide the bulk of both intimate ties and practical support and older people are active as donors as well as recipients of aid in their networks (Phillipson *et al.*, 2004). However, he adds a number of qualifications. Firstly, since many networks are couple rather than family-centred, networks are vulnerable to overload. Secondly, while kinship ties are important, so too are ties with friends, neighbours, co-workers and other organisational ties: 'In reality people place themselves within a complex range of relationships, using different ties for different purposes, but often with limited overlap among them' (Phillipson *et al.*, 2004: 45). We previously discussed Phillipson *et al.*'s findings that kinship and neighbouring ties were strongly affected by broader patterns of social change in particular migratory patterns.

Wenger in her study of the support received by older people in rural north Wales and Liverpool (1989, 1995) identified five different types of network based on three criteria: (1) the availability of local close kin; (2) the level of involvement of family, friends, and neighbours; and (3) the level of interaction with voluntary and community groups (Wenger, 1995: 61–62). The five types were:

1. the local family-dependent support network: mainly relying on close local family ties with few peripheral friends and neighbours;
2. the locally integrated support network: close relationships with local family, friends and neighbours;
3. the local self-contained support network: armslength relationship or infrequent contact with at least one relative and low key community involvement;
4. the wider community-focused support network: active relationships with distant relatives, usually children, and salience of friends and neighbours;
5. the private restricted support network: absence of local kin and minimal contacts with local community. (Wenger, 1995: 62–63).

Wenger also explored the associations between network type and both demographic variables and personal outcomes. It was not surprising that variables such as age, marital status, gender, migration pattern, social class and household composition were correlated with network type (Wenger, 1995: 72–77). Wenger identified two network types as more robust than others in providing informal support – the local family-dependent and the locally integrated. What was especially interesting was that these two network types were more prevalent in Liverpool than in rural Wales which is contrary to the image of strong cohesive community in rural areas and which Wenger explains by differences in population stability (Wenger, 1995: 77–78). We will return to Wenger's typology later when we compare it with the typology of connectedness developed in this study.

Claude Fischer in a study in the San Francisco area examined the structural factors associated with adult personal networks (Fischer, 1982). He found that educational level had the most consistent effect upon personal networks. Other factors included household income and geographical location. Fischer concluded that urban dwellers had more alternatives open to them – social as well as institutional – when developing relationships and could be more selective in their reciprocities with kin than rural dwellers. However, while for higher income households distant friendships could be just as significant as local ones, Fischer concluded that for low income households proximate contacts were often preferred because they could be more easily mobilised. In relation specifically to the social networks of older people a Canadian study concluded that older people who were relatively healthy were more likely to have more diversified support networks available (Keating *et al.*, 2002). Health emerges as an important factor associated with the connectedness of respondents in this study and is discussed further in Chapter Eight.

Summary

In both Rathmore and Rathbeg there was evidence of local extended family where respondents had both children and siblings living in the locality. However, rising house prices have made it less possible for children in Rathmore to live in the same locality as their parents and there were more extensive kinship networks in Rathbeg. Due to emigration and migration the dispersed extended family was a strong feature of the lives of many older people in both localities. Where helping relationships are concerned, as the literature suggests, husbands and daughters appear to form the inner circle of support. To a lesser extent sons and other, usually female relatives also provided practical help. Daughters were the people in whom respondents were most likely to confide, however, husbands and sons were also frequently used as confidants by those who were married and/or had children.

The extent to which older people themselves provide emotional and practical support to others was very evident. Approximately 72 per cent of respondents said that there was someone who relied on them for advice or support. Approximately 20 per cent of co-residing respondents gave help to someone with a specific illness or disability within the household, in many cases this involved full time caring. Approximately 40 per cent of respondents gave regular help to someone outside their own household. While much of this help was given to their own children for example minding grandchildren, help was also frequently and routinely given to neighbours. This involved a range of practical tasks such as doing errands, giving lifts and helping with household chores. It also involved visiting people particularly those whose mobility was restricted in some way.

The vast majority of older people in the study had relationships involving relatives and/or friends that appeared to be satisfying and to involve either visits or social occasions or shared leisure activities. There was however significant variation in the extent to which there was regular contact with relatives, friends or neighbours. Relationships with neighbours could be of immense importance for those who had few relatives and friends. However, there were many people who had extensive kin, friendship *and* neighbour networks in our study areas. The picture concurs with Treacy *et al.*'s findings that the majority of older Irish people in Ireland are not socially isolated and are part of a locally integrated support network that includes family, friends and neighbours, as well as having some degree of participation in church, social groups and clubs (Treacy *et al.*, 2004: 175). In Chapter Eight we go on to develop a typology of the ways in which older people participate in their communities and a model of connectedness.

Chapter Eight
Explaining Connectedness in Rathmore and Rathbeg

Introduction

We have found that interactions within localities through leisure activities, helping activities, voluntary work and informal interactions help to create social and moral worlds where people experience a sense of belonging, involvement and fulfilment. It appeared that where activities were done with others, for purposes of fun or helping or endeavour, these interactions created enclaves of sociality involving mutuality of experiences that contributed to intrinsic satisfaction and promoted connectedness. Such activities and the relationships engendered therein help to create meaning and purpose in the lives of our respondents.

Not surprisingly, we found enormous variation among older people in their social and communal activities and in the nature and level of engagement with others in different contexts. We also found complexities in the processes involved in different forms of sociality, for example the meaning of visiting was far from straightforward. While there is a sense in which each individual has their own personal community and unique configuration of social interactions, the web of relationships and matrix of connectors represent a social reality that pervades geographical localities and intermediate level groups and that can only be understood in cultural terms. We have seen that the intersubjective meanings of social activities and relationships involve sociability, support, solidarity and friendship. The challenge now is how to capture the interpersonal meaning of such communal interactions while attempting to define and measure connectedness in the lives of individual older people.

We wondered whether it was possible to integrate the qualitative and quantitative data on the significant social activities, relationships and commitments of our older respondents. In this chapter we try to develop a composite picture of how older adults in Rathmore and Rathbeg participate in their communities and to explain the elements and dynamics of connectedness. We will present a typology of the principal ways in which older adults live in their communities, keeping the qualitative and quantitative data as disaggregated as possible. We will also present a model that measures connectedness using statistical techniques applied to the survey data. The methods we used to develop the typology and model were as follow. Firstly, we combined relatively simple statistical analysis of our survey data and analysis of the qualitative data. We used this analysis to draw up a typology of connectedness that suggested ten different relational patterns that differentiated our respondents. This enabled us to understand the detail and nuances of lives that were more or less connected within their communities. Secondly, we used more sophisticated statistical techniques that sought to present a more integrated picture of connectedness. This enabled us to develop a model of connectedness and to quantify older people in the study in terms of how connected they were.

Indicators of connectedness

Through our familiarity with the data including the experience of conducting interviews, recording observations, examining the interview schedules and conducting numerous descriptive statistical operations, we were in a position to select five variables that appeared to indicate connectedness in the sense of meaningful communal participation. The analysis that we had conducted to date led us to select indicators related to membership of a club or group, doing voluntary work, informal helping activities, visiting contact with others and enjoyment of collective leisure activities. The five variables selected that indicated connectedness were:

- Whether or not the respondent was a member of a club or group (Q.6a).
- Whether or not the respondent did voluntary work (Q.6b).
- Whether or not the respondent gave help to anybody outside the household (Q.14).
- Whether or not there were events that the respondent did regularly for enjoyment or fun (events), and
- Whether or not a week would go by without the respondent visiting someone or receiving a visitor (Q.26).

Being a member of a club, group or organisation suggests a degree of engagement with communal life. Volunteering is associated with practical social concern for others and promotes positive values of altruism. Informal helping is an expression of reciprocities and empathy with others. All three activities express and give rise to the development of community solidarities and 'other regarding' behaviour. Looking forward to regular enjoyable events is associated with shared fun and enjoyment, and often learning and creativity, and takes account of individual interests and preferences. While visiting or being visited at least once a week is a very basic indicator of contact and support and relates more to one-to-one relationships rather than the quality of relations in a place, it may be especially important for people whose mobility is limited or who tend not to be 'joiners' or who do not wish to take part in group activities. However, we had certain doubts about the validity of visiting contact as an indicator of connectedness. There was an uneven distribution of responses to this question – only one eighth of the sample said that a week would go by without receiving or making a visit. Furthermore, we have already discussed some limitations of regarding visiting as a reliable indicator of connectedness. Therefore we conducted our analysis by selectively leaving in and omitting this particular variable. The frequencies for the index of the 5 items and 4 items are shown at Tables A8.1 and A8.2 in Appendix C.

We used SPSS (Special Package for the Social Sciences) analysis to recode the five variables into dichotomous values based on positive and negative scores and to select subgroups of respondents who were more and less connected on these different dimensions. The subgroups that were analysed through this method were composed of people who had high scores on each of the five dimensions, people who had high scores on four dimensions (excluding Q26), people who had low scores on four or five dimensions and people who had a mixture of high and low scores on different dimensions. We will firstly examine those respondents who were less connected as measured by the indicators of connectedness.

The less connected respondents

Only two of the 165 respondents in Rathmore (1.2 per cent) had low scores on all five indicators, that is they did not do voluntary work, they were not members of any club or group, they did not give help to anyone outside their household, they did not visit or receive a visitor at least once a week and there was no event that they did regularly for enjoyment or fun. When we excluded the question on visiting we found that 16 respondents (9.7 per cent) had low scores on four indicators. The two most unconnected respondents were male, one single, one widowed, and both had worked in unskilled manual employment. Both had significant health problems – one had severe arthritis and the other had heart and circulation problems and neither could walk very far. Both said they experienced loneliness frequently.

Indeed the most striking characteristic of the less connected group was that of poor health as subjectively rated – 13 of the 16 (81.3 per cent) said their health was fair or bad. In the total sample only one third described their health was fair or bad. Approximately two thirds said they had a major illness or disability compared with approximately one half of the total sample. Ability to do daily tasks was also related to being less connected. Half of the less connected group said they were unable to do three or more tasks of daily living. This compares with 14.5 per cent of the total sample. There was a complex relationship between subjective rating of health and objective health. Overall, while there was a close association between subjective health, objective health status and functional ability, the specific ways in which health problems were perceived by the respondents clearly involved subjective elements, perhaps related to expectations, outlooks and how individuals managed their difficulties. Just over two thirds of the less connected group were in the older age group, over 75 years, while just under one third were between 65 and 75 years. In the total sample approximately half were under 75 and half over 75 years.

Respondents who were less connected were less likely to have a major interest, more likely to say they experienced loneliness frequently or often, less likely to say they enjoyed life very much and somewhat less likely to say that other people relied on them for advice or support.[1] Half of the unconnected group were widowed; this was somewhat higher than the proportion (38.2 per cent) of widowed respondents in the total sample. Factors that were not associated with unconnectedness included gender, whether or not they lived alone and the stated importance of religion to them.

The meaning of unconnectedness

However, it would be misleading to suggest that this group whom we have described as less connected are unconnected with their family, have no contact with neighbours or friends or are necessarily uninterested in what happens in the wider community. When we went back and examined the interview schedules of the so-called unconnected group, the picture of their individual lives revealed greater diversity

1 43.8 per cent did not have a major interest compared with 21.8 per cent of the total sample, 56.3 per cent said they experienced loneliness frequently or sometimes compared with 32.3 per cent of the total sample, 12.5 per cent said they enjoyed life very much compared with 52.7 per cent of the total sample, while 62.5 per cent said that someone relied on them for advice or support compared with 73 per cent in the total sample.

and greater complexity in relation to their experiences of sociability and solidarity than the above analysis might suggest. Almost all had at least weekly contact with their families and those who had children said that their children gave them assistance with practical tasks where this was needed. Two respondents lived with daughters and two cared for other household members at home, one for a wife who was ill and one for a son with a learning disability.

What was particularly striking was how many commented positively about their neighbours – eight respondents from the unconnected group spontaneously spoke about friendly and helpful neighbours with comments such as: 'People care about one another here'. There was variation in whether friends and neighbours visit. One widower said he was never a good visitor but he said he had very nice neighbours. Another widower in his 90s had regular visits from family, friends and neighbours and he himself visits regularly. He also said he could talk to friends and neighbours about problems and worries. Yet he said he frequently experienced loneliness and the interviewer described him as 'quite lonely but a happy man.'

Visiting by friends or neighbours may have been inhibited where respondents lived with their children or where the respondent was in poor health or was caring for another household member who was ill or who had a disability. To some extent not visiting among friends and neighbours seemed to be related to poor mobility and coping with illness or disability. In these circumstances social outlets such as clubs or day centres may be all the more needed. Indeed, several of this group commented that there should be more things for older people to do in their area and more day care places. A degree of boredom and loneliness was not surprisingly associated with inactivity and a few respondents appeared open to the idea of taking part in a social or educational activity. One of the two respondents who were caring for someone at home appeared to be very burdened with caring responsibilities and the interviewer described finding a 'level of endurance' in his life. It also appeared that a pattern of regular visiting tended to happen less in some neighbourhoods than in others. In one or two cases negative perceptions were expressed about the neighbourhood in relation to vandalism and a lack of neighbourliness.

While having and valuing good neighbours was common among this less connected group, not all the respondents regarded regular interaction with neighbours as important. One respondent, a widow, was described by the interviewer as 'having no interests outside her children and grandchildren', another widow was described as 'intelligent, plain spoken, independent, not into group activity', while a third, a married woman, was 'not very involved in the community and likes it that way.' The widow who was described as independent was particularly interesting in that, while she had little interest in social activities in her locality and did not find her neighbourhood friendly nor did she value general friendliness, was very interested in politics and current affairs. She commented: 'You'd be dead and buried before anyone would know ... people keep to themselves ... that's urban life.' However, in response to the question about the importance of being greeted, she replied that it was not important and added 'it wouldn't bother me if they didn't.' It appeared that general life circumstances had made her independent and self-sufficient with many solitary interests such as reading, writing and crochet. She had gone back to study in her late 50s and she expressed an interest in finding out about adult education classes and in learning to use a computer. She was perceptive about the experiences of ageing, commenting:

> Ta a reidh beigneach caite.[2] You become attuned to reclusivity. Your main aim is to maintain your independence. Your main wish is never to be a burden on anyone. Health is number one.

She had also very definite views on how older people could be more involved in their communities. She suggested that schools and churches had space that could be utilised at different times. She added: 'All their life experience and wisdom is not tapped. Opinions of older people are not listened to – they're shoved to one side.' This lady could hardly be described as passive or uninterested in the general community, yet she did not express this in sociability or solidarity with people in her immediate locality.

2 Speaking in Gaeilge this means 'Their time is nearly spent'.

Another widow who had no children also presented a self-sufficient and positive outlook. She said she did not experience loneliness adding: 'I'm self-sufficient in that manner. I have my books and reading and say my prayers. What more do I want?' She went to a day centre weekly and got good support from the health and social services which she expressed herself satisfied with. This is an example of services meeting social needs that did not involve too much intrusiveness given this lady's desire for privacy and independence.

The lives of the two least connected respondents reveal that even with some family and social services support they experienced considerable loneliness and lack of fulfilment in their lives. The first man, a widower expressed himself happy with the support he received. He has one son who calls at least once a week and who brings him for hospital appointments, and nieces who call and cook for him and whom he described as 'very, very good'. A care attendant calls four times a week and meals on wheels are delivered to him three times a week. He said that old people are looked after 'better than ever'. As regards having someone to talk to he said he could have a chat with his son and one or two others. He described how he spends his time: 'Watching people doing things during the day and then I watch television.' He described his loneliness: 'Loneliness is worse than arthritis, I've been lonely, all the lads I know drifted away. To live on your own is the world's worst.' The interviewer described him as 'very lonely and suffering a little from forgetfulness.' The interviewer said that he wanted to keep him as long as possible, that he was there for over two hours and that the respondent enjoyed the chat.

The second of the least connected respondents gave some details about his earlier life. He had emigrated to England for work when he was young but had to retire in his 30s because of an accident. He described the loneliness he experienced vividly:

> Loneliness is a disease in itself. There's no cure for it. You can't bore other people with your complaints. Everyone has their own cross. You have no one. People write you off if you can't do things. They don't want to be bothered with you. I copped on to that very quickly.

This man said that he chose not to go to the various social activities that were available where he lived because of his health problems: 'I don't go into the events. I know that I'm welcome but I'm embarrassed with my complaints.' He also said that he had had some 'nerve trouble'. He referred to social events in a nearby community hall but added: 'It's more or less for women … I don't go in for things like that.' He said he regretted that he could not travel now due to poor health. He also spoke about a past interest in playing music. When the weather is good he walks around the green and sits on the bench. He found that 'people are not that friendly' and that 'they won't stand up and chat.' In reply to the question on having someone to confide in he said that there was no one with whom he can discuss problems or worries. His brother brings him to his house for dinner once a week but his brother travels a lot and he might not see him for a few weeks at a time. His sister-in-law calls to do some household chores for him once every two weeks. Apart from these relatives the only other visitor he has is the Priest who brings Communion every month. He was less satisfied with the social services support he received as the following comment illustrates:

> I think the Public Health nurse should call more often … she wants me to go to the clinic but I told her she has a car. She was annoyed about that … I had to walk to the chemist and doctor. She should call more often. [I said] 'You're drinking cups of tea and talking when you should be here and you want me to walk there with two sore legs.'

He also complained about a long wait for a walking frame and rails to get in and out of bed. He said he used to get meals three times a week from the meals on wheels service but that this had been reduced to two. He was not satisfied that he had to wait for 6 months for a heart operation which was 'too long'.

The author noted that while at the beginning of the interview this respondent seemed wary, he then opened up and appeared to enjoy the interview. At the same time he may also give the impression of being somewhat contrary and inclined to complain a lot.

The more connected respondents

When we selected people who were connected on each of the five indicators we found a total of 17 respondents (10.3 per cent of the total sample).[3] These were people who did voluntary work, were members of a club or organisation, gave help to someone outside their own household, visited or had a visitor at least once a week and said that there were events or activities that they did regularly for enjoyment or fun. We found that similar factors associated with unconnectedness, such as health and age, were associated with being more connected. Respondents almost unanimously described their health as good or very good (16 out of 17). This was despite just over one half of this group reporting that they had some major illness or disability.[4] Approximately three quarters were in the younger (under 75) age group while one quarter were in the older (over 75) age group. Over 80 per cent of the more connected group could do all daily tasks compared with just under 60 per cent of the total population. Therefore positive views of one's health, functional ability and being under 75 were characteristics associated with being connected.

All 17 (100 per cent) of the most connected group had a major interest compared with 78 per cent per cent of the total sample. The major interests of this group included hill walking, golf, music, gardening and politics. Furthermore, all but one of the 17 (94.1 per cent) said that religion was very important to them compared with 69.7 per cent of the total sample. Thus it appeared that attaching high importance to religion was positively associated with connectedness. However as we noted previously importance of religion was not associated with unconnectedness. For the most connected group it is possible that their sense of the importance of religion was also associated with social engagement through church groups. Many of them were involved in church activities such as being members of a Parish Council or

---

3 The same proportion and the exact same respondents were included when we omitted Q26 and used 4 indicators.

4 In the total sample two thirds described their health as good or very good while just under a half said they had a major illness or disability.

acting as a Eucharistic Minister. It was interesting to note that almost 65 per cent of this group were involved in two or more clubs or organisations compared with just over a quarter of the total sample who were involved in two or more groups. The most connected were more likely to belong to the white-collar classes while the less connected were more likely to belong to the manual classes. Approximately three quarters of the most connected group belonged to white-collar backgrounds while one quarter of this group belonged to manual backgrounds.

Enjoyment of life was very evident among the most connected group – the overwhelming proportion, 88.2 per cent, said they enjoyed life very much, 11.8 per cent said they enjoyed life quite a lot while none said they enjoyed life only sometimes. In contrast, 52.7 per cent of the total sample said they enjoyed life very much, while only 12.5 per cent of the less connected group said that they enjoyed life very much. Respondents in the most connected group were somewhat more likely to say that other people relied on them for advice or support; 76.5 per cent said that others relied on them compared with 73 per cent in the total sample and 62.5 per cent of the least connected group. None of the most connected respondents said they experienced loneliness frequently while under one quarter of them said they experienced loneliness sometimes.[5] Neither marital status, gender nor whether the respondent lived alone or co-resided appeared to be related to connectedness.

Those who were most connected engaged in a wide variety of interests and social activities that included hobbies, voluntary work and family commitments and they appeared to be contented with their lives. For example one married lady was actively involved in party politics, she was a Minister of the Eucharist in her local church, she looked after her grandchildren regularly as well as an older aunt and she was involved in visiting friends and neighbours on a regular basis. A married man was involved in the St Vincent de Paul Society and an

5 Among the total sample only 5.5 per cent said they were frequently lonely while 26.7 per cent said they were lonely sometimes. Among the unconnected group we saw that 56.3 per cent said they experienced loneliness frequently or sometimes.

Active Retirement Association. He brought his grandchildren to school every morning and walked every day in the local park. He played a musical instrument, enjoyed DIY and regular outings to the pub.

Many of this group were physically active in sport and many were interested in travel and spoke about forthcoming holidays. Several of these respondents came across as having a good sense of humour and/or enjoying life. This was the description that one of the interviewers gave of a single lady who had many interests and commitments: 'Very vivacious, good sense of humour about old age, great network of friends. Helps others. Interested in further education.' This lady helped her sister and four other friends/neighbours on a regular basis. She enjoyed art classes, gardening, reading, crosswords and going to the pub with her friend. She regularly chatted with her sister, friends and neighbours and said that if she was in the front garden 'she wouldn't have a minute.' She said that everything she did was for enjoyment. Another interviewer described a male respondent as: 'Very active and willing, enjoys life and having the time to do the things he enjoys.' This man helped in a sports club, did church work, visited and helped a neighbour with gardening and had a major interest in hill walking.

A number of the most connected respondents demonstrated altruism and practical social concern through their voluntary work in their community. For example two residents of sheltered housing complexes worked consistently through the Residents' Association to improve the general physical and social environment of the complex. They demonstrated an awareness of the needs of others, they had a strong religious commitment and they derived satisfaction from their voluntary work. It was interesting that in both cases their evident social concern seemed to have emerged from difficult personal circumstances. One of them gave insight into his philosophy of life through the following comments:

> All life is fun and enjoyable … the possibilities of the day which lie ahead.
> It's important to make eye contact … it's especially important for older people to say hello to them.
> All you can do is show good example and be cheerful.
> I'm at the service of people if they need to go to the hospital, hardware shop etc.

His good work was appreciated by other neighbours, several of whom commented spontaneously on all the work he did and on his general helpfulness.

The other socially committed resident speaking about particular difficulties that she had to overcome in her life said: 'There is satisfaction in struggle, you do achieve a lot.' She had done voluntary work for many years, had participated in adult education activities, was active in an ARA group and she said that her religion was 'something that was very important, something that was comforting.' She spoke about the social concern people had for each other in the complex: 'We will all help if someone is sick. We all watch out for one another.' She was concerned about residents who did not want to take part in the social activities that the Residents' Association organised: 'There's a little pocket of people in the complex who are reclusive, a lot are alcoholic … If people don't want to take part you have to respect their privacy.'

Another 'connected' respondent said she was motivated by 'a strong religious faith'. She was very involved in her church – in the Ecumenical Movement, as a member of the Pastoral Council, and in fund raising for Missionary activity and refugees. She regularly visited an older person in a nursing home, brought food to neighbours if they were sick and was involved in daily visits to friends and neighbours. In addition she offers hospitality to a traveller lady[6] who calls to her house twice a week. She described her faith along with gardening as her major interests. For our most connected respondents their social and helping activities were often intertwined with a strong religious and humanitarian commitment.

However, it is not necessary to be a social activist to be connected. People who took part in organised social activities often preferred a passive rather than an active role. Many people who were not involved in groups of any kind or in volunteering had close supportive relationships with neighbours and friends. We identified people who were good neighbours in the small but helpful things they did for each other. They may not have wanted to socialise with neighbours on a regular basis but they did errands and chores for neighbours. Even

6 A member of the ethnic minority Traveller community in Ireland.

those who were frail were involved in small but meaningful reciprocities with neighbours. Other forms of altruism included fundraising and giving money to charity. Giving money to support Irish Missionary work in developing countries was a form of connectedness that was common in Ireland in the past. This practice continues albeit much of it is now given in the form of development aid. For older people who could no longer be active in their church or in organisations, this form of giving, particularly to religious orders working abroad, allowed them to continue to express social solidarity. We cited the example of the lady in Rathbeg who spent large amounts of her time knitting jumpers for children in developing countries.

Connected on some dimensions

The diversity of modes of connectedness is further underlined when we consider respondents who had high connectedness scores on some indicators and low scores on others. We examined the lives of respondents who had high scores on four indicators. The common indicators were doing voluntary work, being a member of a club or organisation and being involved in regular visiting. In the case of one group they also helped someone outside the household but said there was no regular activity they did for fun or enjoyment. A second group did not give explicit help to someone outside their household but they took part in regular enjoyable events.

The first group was a small group, overwhelmingly female, who were very involved in their communities in a helping capacity. One was a nun recently returned to Ireland who did voluntary counselling work in the area. Another respondent looked after her autistic daughter in addition to doing voluntary work and helping a friend with many tasks of daily living. A third respondent, aged 65 and a full-time homemaker, stated that 'she worked all her life unpaid' and that she was 'always being approached by friends and neighbours'. Elaborating on her response that there were no events that she did regularly for enjoyment or fun she added that 'her day was so full.' The interviewer commented that 'the interviewee is still a full-time mother and home-maker, the category of older person does not apply to her.' In the case of these individuals we can see that their commitments and circumstances may have prevented them from developing and participating in hobbies and enjoyable social activities.

The second group was a larger group than the previous group and had more males than females. Having a major interest was a strong characteristic of this group. While there was great variety in respondents' interests and leisure activities, there were a number of common interests and commitments, notably involvement in sport, church, special interest groups or Residents' Associations. Voluntary work was often connected with an interest in sport – for example being actively involved in a GAA club, golf club or bowling club. Their voluntary work often involved charitable or community work and their interests included creative activities and adult education.

This group may not have been called on to give explicit help to someone outside their household and while most communal involvement was locally based, there were a few respondents who had little involvement with people in their own locality. One respondent was active in voluntary work through his life-long interest in and achievements in a particular sport but this was not confined to his own locality. Indeed he commented: 'I don't tend to focus on the community.' His community of interest was the sporting community. He led a busy life in that he still did part-time paid work as well as voluntary work, he socialised regularly with friends in a club and he had close relationships with his family and siblings. An interesting comment he made was that he did not like the idea of growing old and the interviewer recorded that he still lives his former lifestyle as much as possible. However, for the vast majority of this group most of their voluntary and leisure activities occurred in their general locality and these respondents tended to comment favourably on their neighbours and neighbourhood. One respondent who led a very active life including doing some paid work, voluntary work and pursuing hobbies spoke about people in the area being 'so kind, considerate and helpful' and suggested that greeting people on the street 'was what community is all about.'

Ten relational types

Having analysed the experiences and characteristics of respondents who participated in their communities in different ways and who have been defined as more or less connected along significant dimensions, we will now attempt to draw up a typology of the main ways in which older adults in Rathmore and Rathbeg are connected within their communities. Therefore, while the relational patterns discussed below correspond closely with the characteristics of the groups identified as being more or less connected, it is not possible nor theoretically valid to state categorically how many respondents belong to each type. The ten relational patterns do not constitute distinct or separate groups; one type may overlap or merge with another type. Rather they suggest broad orientations that take account of levels and degree of contact with family, kin, neighbours and friends, as well as participation with others in their community through leisure, voluntary work and informal helping. The ten relational types are shown below:

1. Active life
2. Extended middle life
3. Supportive people in dense networks
4. Small number of close relationships in neighbourhood
5. Family oriented
6. Family and interest oriented
7. Participants in traditional activities for older people
8. Self-sufficient and independent
9. Restricted by poor health
10. Less connected

1. Active life:
The 'active life' group refers to people who are involved in a wide and dense network of relationships. Such relationships include supportive helping relationships with family, kin, neighbours and friends as well as relationships related to their voluntary work and leisure activities. They appear to derive intrinsic satisfaction from all their interests,

commitments and relationships. Some may be still doing paid work on a part-time basis but if they are retired they are likely to say that they are enjoying retirement very much. Many of them are typically involved in more than one type of voluntary/community work. They are frequently used as confidants or advisers by other people and their voluntary work is generally valued. Included among this type are people who demonstrate through their work and through their language an altruistic outlook and frequently a strong Christian ethos. They see their work as trying to improve the lives of others in their locality. Their interactions would typically bring them into contact with a range of different age groups and they usually have good social and organisational skills.

2. Extended middle life:
These are people who are living in nuclear family units with their adult children still living at home. The women are still full-time homemakers and the men are recently retired or partially retired. They are typically in the younger age bracket, are often active in sport or Active Retirement groups and have specific hobbies and interests. Some of them do voluntary work which may relate to a hobby or church membership. They do not see themselves nor do others perceive them as 'old'. They have not experienced any major transitions associated with ageing and their focus is on family and interests. As they move through life transitions and when their adult children leave home they would most likely be classified in types 5 or 6.

3. Supportive people in dense networks:
This category refers to people who are involved in supportive helping relationships with family, kin, neighbours and friends. They have strong friendships and give help and support to others. Individuals in this category may belong to any marital category or age group and they are not restricted by disability or poor health. They frequently visit the homes of neighbours and friends who may be restricted in their mobility and to whom they render practical assistance. Traditional volunteers are associated with this category, for example people who have worked or who still work in voluntary services such as meals on wheels, church groups, day centres or Society of St Vincent de Paul.

For women especially, close friendships have developed with other volunteers and this work is an important part of their lives. Some respondents who care for a disabled son or daughter also find time to give their services voluntarily to a disability organisation. However, the demands of caring work can also limit opportunities to pursue hobbies and other types of social interaction.

4. Small number of close relationships in neighbourhood:
This category refers to people who do not generally have a wide network of relatives or social contacts, are usually single or widowed and tend to live alone. They are, however, close to one or two neighbours, are frequently in each other's homes and give each other practical help and support. Older people who are in poor health or who are frail or incapacitated will be unable to do return favours but they do show appreciation of the help given and value the friendships involved. People of this type may participate in traditional social activities within their locality such as bingo and social outings, preferring more passive forms of leisure and social involvement. Their friendships can be long-standing or more recent but appear to be based on common understandings and shared experiences. Such friendships were observed among women living in sheltered housing where there was a sense of sharing life space. On the other hand there were respondents who did not wish for this type of friendship with neighbours and who would have regarded closeness of this nature as intrusive (See category 8 for example).

5. Family oriented:
This type comprises individuals who are very involved with their children and grandchildren. For women their interests are often centred on domestic work including cooking and craft work. They derive much satisfaction from organising or participating in family occasions. In the case of some married couples in this category husband and wife often socialise in different settings, for example the husband will go to the pub for a drink while the wife may not. Some women said that they missed their children when they set up independent households. Many respondents spoke about the loneliness they experienced as a result of widowhood and in this situation many depend on their

children for practical support and companionship. While clubs such as Ladies' clubs can be an important social outlet for widows and the pub plays a similar role for men, many are not 'joiners' and would be shy in a group situation. The telephone plays a very important role especially for women who spend a lot of time at home. In some situations of illness or incapacity caring can be burdensome for a spouse and can limit other social interaction.

6. Family and interest oriented:
This category includes people who are very involved with their families and who have particular hobbies and interests. Their interests may include sport, a craft pursuit or a hobby that developed over a period of time and that may also be the basis of friendships. Social activities are family centred and interest based, for example they may revolve around a sports club such as a golf club. Their leisure and social activities are not necessarily local and may involve a wider network of friends. This category appears to apply particularly to men and is more common among middle class than working class respondents.

7. Participants in traditional social activities for older people:
This category includes people who participate in traditional activities associated with older people. They enjoy social activities such as bingo, sing songs and organised social events/outings. Most people in this category are female but they may be married, single or widowed. They enjoy the company of their peers and many have been involved in Senior Citizens' clubs or Ladies' clubs over many years. Most prefer not to take an active role in the group. The friendships made are very important and can be close and/or companionable. Active Retirement groups have become more popular in recent years and cater for a broader cross section of older peoplc, including men. However, some ARAs also cater for people who wish to be involved in a more passive way and who prefer traditional activities.

8. Self-sufficient and independent:
This category refers to people who are very self-sufficient and have individual, usually home-based interests. Being involved with others

either in the locality or through group activities is not important to them. People belonging to this category were found in both the younger age group and the older age group. They tend to have most regular contact with close family members. Being able to continue to live independently is very important for the majority of older people but particularly for this group who may fear losing their independence. They desire social support and social interaction without intrusiveness. Forms of social support may or may not be acceptable such as a home help service or a day centre. However, having local access to cultural/social amenities such as a library, post office, pub, church and having accessible transport are particularly important for older people and enable respondents such as these to pursue their interests and in this way to continue to participate in their communities.

9. Restricted by poor health:
This category refers to people whose health prevents them or restricts them from taking part in social activities. They may have been more actively involved in social activities previously but they now tend to spend a lot of their time in the house. Many are involved with their family and some live with their children. Many have contact with neighbours and value their good neighbours. They are mostly content with their lives but get bored or fed up sometimes and many would like to have more opportunities to meet other people. Day centres and local social activities can be an important outlet for this group. However, if there is no one to encourage or to accompany them to a club or activity they may remain somewhat socially isolated.

10. Less connected people:
This category contains people who have few social contacts and limited social activities. They may experience poor health and limited mobility and may have had health problems at an earlier stage in life. They often experience loneliness and some may be shy or may not have good social skills. Some have become somewhat reclusive and are perceived by others as not wishing to help themselves by availing of opportunities for social engagement. Others perceive this group as having all the greater need for social contact. This group also includes male respondents who do not have regular social contact with their

children or with all their children. In many cases this has arisen due to marital separation. Marital separation and lack of a steady work career may have put people in a difficult financial situation and further limited their opportunities to pursue hobbies or engage in social activities. A small number of men appear to be heavy drinkers and they spend a large part of the day in the pub. While we argued previously that the pub is an important social venue and outlet especially for men who live alone, it may also become a refuge from a life where there are few meaningful relationships. Less connected people are more likely to be male, to live alone and to belong to the manual classes. Some find they have too much time on their hands.

The Typology: modes of participation

The typology takes account of types and levels of involvement in purposive and enjoyable interests and social activities, and the different ways in which older people contribute to the lives of others. It also takes account of health and mobility as well as personal life circumstances. Each category represents a particular communal orientation that distinguished our respondents and that can be empirically verified both from the quantitative and qualitative data. However, it is important to point out that these relational patterns do not constitute distinct or separate groups; one type may overlap or merge with another type.

For example people defined as living an 'active life' are involved in dense social networks involving kin, friends, neighbours and the wider community and derive satisfaction from a wide range of interests and social activities, often taking a leadership role. People described as supportive people in dense networks also derive considerable satisfaction from their helping activities, and are frequently the backbone of many traditional voluntary groups. However, they do not necessarily have consuming hobbies or take part in social activities, and their work is often low-key and informal. In contrast people who have a 'small number of close relationships in the community' derive companionship, satisfaction and support from these friendships with neighbours and prefer passive forms of leisure. Some categories may merge. For example some individuals in the 'extended middle life' category could also be classified as having 'a dense network of one-to-one relationships' if they are also involved in supportive relationships outside the family. Participants in 'traditional social activities for older people' may overlap with those who have a 'small number of close relationships in the community'.

While the analysis involved in constructing the typology does not allow us to quantify the number of respondents in each category, we can say that the 'active life' type corresponds with the 10.3 per cent of the sample who were found to be connected on all five dimensions. Those respondents who were unconnected on all five dimensions (two individuals or 1.2 per cent of the sample) would be found in category 10, 'the less connected' type. The 9.7 per cent who were unconnected on four aspects can be found among categories 8, 9 and 10. They include people who had fewer significant family or non-kin relationships, those who were restricted by poor health and those who were not involved in many networks of sociality partly at least by choice. The number of respondents in categories 2 to 7 cannot be quantified from the analysis.

The typology suggests that there are many different ways of being connected and of participating in one's community as an older person. While having dense and wide relationships with kin and non-kin and being involved in meaningful communal activities is associated with satisfaction in later life, as we already pointed out it is not necessary to be a community leader or a dynamic volunteer or to be friends with lots of people in one's locality to be connected. The idea of enclaves of sociality that pervades the typology applies to more passive as well as more active forms of participation and to people who are primarily family focused or who have a small number of friendships. The importance of the contribution that older people themselves make to communal life is underlined by the typology. The informal helping activities, communal leisure, involvement in church and groups and generalised friendliness that were widespread among our respondents illustrate the interdependencies and reciprocities that bind older people with kin, friends and neighbours and that give meaning and satisfaction in life.

The importance of positive neighbouring is highlighted for most of the categories. The availability of different types of groups and the possibility of different levels of involvement – more passive or more active – create opportunities for older people to participate in ways that suit their own predispositions and interests. Such groups also provide exemplars for positive sociality including altruism and practical kindness. The typology suggests that forms of connectedness associated with face-to-face relationships with others in one's locality promote both personal and social well-being. Therefore, we argue that the meaning of community as a social place within which multidimensional relationships can be experienced is pivotal to obtaining satisfaction in one's older years.

The typology is entirely different from that developed by Wenger to measure support networks of older people. Wenger's typology is based on the availability of family and kin and on rather static measures of interaction with significant others, such as attending religious or club meetings or chatting with neighbours. Wenger herself acknowledges that in her analysis the formation of support networks is 'largely outside the control of the individual, depending largely on the experience and action of others' (Wenger, 1995: 62). It thus depicts a somewhat passive role for the older person in terms of their support needs and results in a rather general classification of older people related largely to socio-demographic characteristics. In contrast the typology of connectedness developed in this study is more dynamic and interactional.

Model of connectedness

Applying more sophisticated statistical techniques,[7] we sought to develop a model of connectedness based on the five indicators of connectedness. The statistical analysis used is described in Appendix D. While we lose sight of the nuances of the different modes of communal participation, this analysis enabled us to quantify older people

7 Latent class cluster analysis.

in terms of how connected they were and to seek to explain the characteristics of people who were more and less connected.

The statistical analysis showed that two models fit the survey data; when five variables were used 3 clusters emerged and when four variables were used 2 clusters emerged.[8] The 3 clusters that emerged when five variables were used (Model 1) were:

1. Limited level of connectedness
2. High level of connectedness, and
3. High connectedness through voluntary work/low level of social involvement

The 2 clusters that emerged when four variables were used (Model 2) were:

1. Limited level of connectedness, and
2. High level of connectedness

We show below the number of respondents grouped in each of the categories according to Model 1 and Model 2.

Model 1

Limited level of connectedness	81 respondents
High level of connectedness	74 respondents
High connectedness through voluntary work/low level of social involvement	10 respondents

Model 2

Limited level of connectedness	73 respondents
High level of connectedness	90 respondents
(2 cases with missing values)	

8 The five variables were: q6a, q6b, q14, q26 and events. The four variables were: q6a, q6b, q14 and events. Q 26 was excluded.

Model 1, the 3 cluster analysis, highlights a small group of ten respondents who are very involved in voluntary/community work but not necessarily involved in other types of social activities – for example they did not take part in regular enjoyable events and many of them were not involved in regular visiting. This appears to identify respondents who were very committed to helping activities but did not engage in communal leisure or other types of social interactions in their locality.

Factors related to connectedness

We carried out crosstabulations of the key independent variables with the 2 and 3 model clusters. Table 8.1 summarises the correlations between the independent variables and the two models, using Pearson chi-square. In relation to Model 1 the following variables were significantly correlated with connectedness: *subjective rating of health, functional ability, having a major interest, age* and *whether or not there was someone who relied on them.* Having a major illness or disability was not statistically significant. In relation to *enjoyment of life, importance of religion, experience of loneliness, social class and educational level*, while these variables were not significant at the .05 level, there was a strong association between them and being connected. Those in categories 2 and 3 (high level of connectedness and high level of voluntary work/low level of social involvement) were much more likely to say that they enjoyed life very much, to state that religion was very important and to belong to a white-collar background. While absence of loneliness was not statistically significant, there was a tendency for those who were connected and especially those in category 3 to say that they never or rarely experienced loneliness. Respondents who had more than Primary level education were more likely to belong to the connected group (either category 2 or 3). Marital status, gender and being co-resident or living alone were not significantly related to connectedness. However, respondents who belonged to category 3 were predominantly female (8 out of 10) and were more likely to be co-resident.

When we did the same crosstabulations with Model 2 we found that there were similar patterns (Table 8.1). Enjoyment of life and importance of religion both had stronger association with connectedness, however they remained below the .05 level. Another interesting finding was that receiving regular visits from relatives, friends and neighbours were each separately significantly related to connectedness as measured in Model 1. However, in Model 2 visits from relatives and from neighbours were not significantly related to connectedness. However, having visits from friends was significant for both models. The importance of friendship in understanding connectedness is suggested by this finding and we will explore this further in Chapter Nine.

Table 8.1: Crosstabulation of independent variables with Models 1 and 2

	Model 1 (three clusters)		Model 2 (two clusters)	
Independent variables	Chi-square value	Significance	Chi-square value	Significance
Rating of own health	17.1	.000	15.3	.000
Having a major interest	13.8	.001	8.5	.004
Age group	11.4	.003	8.3	.004
Functional ability	12.8	.01	14.6	.001
Saying that others relied on them	6.6	.04	3.9	.05
Enjoyment of life	7.6	.11	5.1	.08
Occupational status	4.4	.11	2.6	.11
Importance of religion or spirituality	6.9	.14	4.8	.09
Gender	2.7	.26	0.07	.79
Having a major illness	2.4	.3	0.28	.6
Marital status	4.6	.33	1.1	.57
Educational level	8.8	.36	7.5	.11
Living alone or co-residence	1.5	.48	0.05	.82
Experience loneliness	3.2	.5	1.4	.5

The diagram below is a visual representation of the models of connectedness. Model 1 is represented by the outer circle, Model 2 is represented by the oval, while the group who scored high for helping activities and low for social involvement (category 3 of Model 1) are represented by the oval with the broken lines. The factors that are significantly related to connectedness are immediately outside the circle while factors that are related but not as statistically significant are shown below this with the arrows. Factors that appear to have no relationship with connectedness are shown in the box.

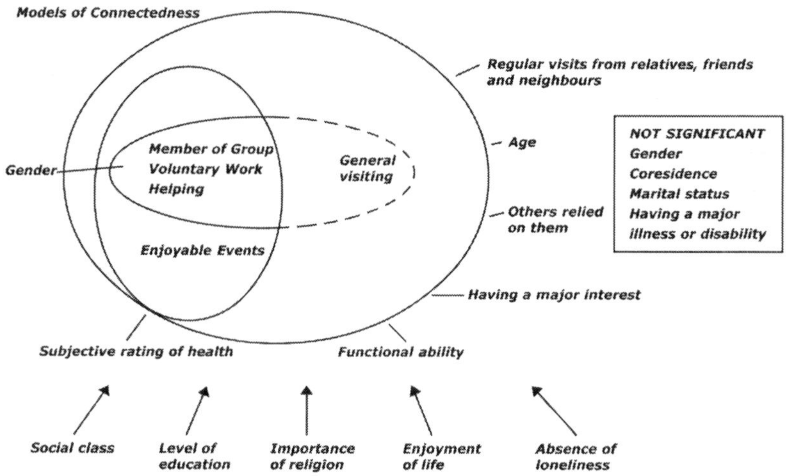

In terms of how we measured connectedness older people who are healthy, functionally active and in the younger-old age group are more likely to belong to the connected group. Those who are classified as less connected through the statistical analyses include people who are simply too ill or too frail to give practical help to others or to attend social events. This does not necessarily mean that they are unconnected with or not involved in satisfying relationships with family, friends and neighbours. Therefore, the typology of relational patterns may offer a more rounded account of the different ways of being connected that is not as activity based as the model of connectedness suggests.

The analyses we have undertaken of the relationships and interactions among older people in Rathmore and Rathbeg suggest a particular theoretical understanding of community. The significance of the five indicators of connectedness seems to be that they capture key elements associated with 'community-ness' as Calhoun defined it – face-to-face interaction, commonality of purpose, familiarity and dependability. There is also a strong moral element suggested in people's own understanding of communal engagement. The significance of friendship is also highlighted in the analyses. It is associated with connectedness as measured by either Model 1 or Model 2. Friendship is also a strong feature of the lives of the vast majority of our respondents as represented in the ten relational patterns. Friendship appears to be strongly contextualised in the daily interactions of older people in Rathmore and Rathbeg through habitual social engagement and helping activities in clubs, church, voluntary groups and other public venues such as pubs and parks.

Summary

This chapter sought to integrate the qualitative and quantitative data on communal interactions of older people in Rathmore and Rathbeg. Connectedness was defined in terms of what appeared important to older people themselves, what resonated with the literature and what patterns of participation occurred. Five indicators of connectedness were selected that suggested meaningful communal engagement. These were: being a member of a club, group or organisation, volunteering, helping others informally, looking forward to regular enjoyable events and visiting or being visited at least once a week.

Based on respondents' scores on the indicators we identified groups of respondents who were most connected, least connected and those who had high scores on some indicators and low scores on other indicators. We described the lives of people who were strongly connected – the wide variety of interests and social activities they had, their extensive helping activities, both through formal volunteering and informally, the degree to which they enjoyed life and the extent to which many appeared to be motivated by their religious beliefs. We also noted that they tended to enjoy good health, to belong to the white-collar class and to be more likely to be in the younger age group.

In describing the lives of those who were least connected, we found that there was diversity and complexity in relation to their circumstances and their intersubjective experiences. What they had in common, however, was that only a very small proportion of them said they enjoyed life very much, a high proportion said that their health was poor and over half said they experienced loneliness sometimes or often. While health difficulties and functional incapacities created barriers to social integration, in some cases it could be inferred that cumulative life circumstances had created difficulties and obstacles to forming satisfying relationships. However, those whom we classified as 'unconnected' were not necessarily socially isolated nor dissatisfied with their lives. Many enjoyed close family relationships and valued what they regarded as good neighbours. Some were independent and chose not to become involved in group activities, preferring more solitary pursuits. However, boredom was frequently experienced by those who had restricted mobility.

A typology of ten relational patterns was developed that pointed to different communal orientations and that differentiated the respondents in terms of how they participated in their communities. The typology is based on respondents' social activities, significant relationships and group affiliations across a wide range of arenas and reflects the different ways in which older people contribute to the lives of others, whether family, other kin, friends or neighbours. It also took into account their living circumstances – stage of the family lifecycle, whether the respondent was in the 'third' or 'fourth' age and their health. It can, therefore, be seen as a measure of both how 'vitally involved' our respondents were and their everyday interactions framed within the socio-cultural parameters of their lives.

The typology suggests that there are many different ways of being connected and of participating in one's community as an older person. While having dense and wide relationships with kin and non-kin and being involved in meaningful communal activities is associated with satisfaction in later life, connectedness does not necessarily mean being a community activist nor having lots of friends. The idea of enclaves of sociality suggests that sociality resides in different groupings and in a variety of settings, from family/kinship groups to church groups to clubs or special interest groups. Meaningful communal interaction includes more passive forms of participation, family focused activities and social/supportive interactions involving a small number of friends. The informal helping activities, communal leisure, involvement in church and groups and generalised friendliness that were widespread among our respondents illustrate the interdependencies and reciprocities that bind older people with kin, friends and neighbours and that give meaning and satisfaction to life.

A model of connectedness was developed based on the key indicators that measured and quantified the degree of connectedness of the respondents. Model 1 was developed from five variables and Model 2 from four variables, omitting the variable related to visiting contact. Using Model 1 we found that 45 per cent of respondents could be regarded as connected, 49 per cent were unconnected and 6 per cent were connected through voluntary/helping activities but not through leisure/social activities. Using Model 2 we found two groupings, 55 per cent could be classified as connected while 45 per cent were classified as unconnected.

The factors that were found to be significantly related to connectedness were: subjective rating of health, functional ability, having a major interest, age and whether or not there was someone who relied on them. There was a strong though not statistically significant correlation between connectedness and the following factors: enjoyment of life, importance of religion, experience of loneliness, social class and educational level. Marital status, gender and being co-resident or living alone were not significantly related to connectedness. However, the small group who were very involved in voluntary/helping work but not necessarily involved in other types of social activities were predominantly female. Having visits from friends was significantly related to connectedness as measured by Model 1 and Model 2. Friendship also emerged as an important aspect of the lives of the majority of older people who were connected in different ways and to different extents.

Chapter Nine
Community and Friendship

Introduction

The theme of this chapter is friendship. Our particular interest is in friendship in socio-spatial settings. Friendship in Western society is seen as a voluntary relationship between peers and we have become accustomed to thinking of friendship as a personal attachment to people who are not socially defined as kin (O'Connor, 1998: 122). However, we encountered friendship in a variety of settings in Rathmore and Rathbeg and significant friendships were not confined to people's personal worlds. We suggest that friendship among older people is best seen as a relational concept characterised by different processes in different contexts including intimacy, solidarity, sociability, respect and acknowledgement. We argue that friendship is a key element in understanding what forms of social connectedness translate into good communities where older people are well integrated, live full lives and have opportunities to contribute to their communities.

We have already documented the extent and patterns of informal help and support that people give each other in daily life including helping relationships among kin and neighbours as well as supportive dyadic relationships. We have also described a pervasive atmosphere of kindness and friendliness among many of the social groupings in Rathmore and Rathbeg. We do not propose to introduce new data in this chapter but rather to examine aspects of friendship in more depth, in particular sociability, affirmation and obligation. In relating our own empirical findings to the literature on friendship we argue that some accounts of friendship in contemporary society may overemphasise psychological closeness as a defining feature of friendship and underestimate both sense of obligation and development of shared moral understandings among friends. Accounts of friendship that emphasise choice and intimacy for example do not fully reflect friendship bonds among neighbours, members of groups and among kin. Furthermore, they do not take sufficient account of elements such as respect, common values and shared experiences. We question assumptions that there are fixed boundaries between the categories of relatives, friends and neighbours. We also discuss the significance of a generalised form of friendship involving habitual cultural practices that are constitutive of identity and well-being. The focus of this chapter, therefore, is to examine the extent to which friendship networks in both formal and informal groupings may create social worlds that make life more habitable, give meaning to lives and create enclaves of sociality where personal and social well-being are sought and attained.

The concept of friendship

In seeking to understand the contexts, forms and significance of friendships among our respondents we examine perspectives on friendship in the work of Raymond Pahl, Graham Allan and Pat O'Connor. Pahl has argued that friendship based on choice and mutual support creates qualitatively different types of relationships in contemporary societies and potentially new forms of collective identities. Pahl has suggested that friends are taking over from families as our major support group in both emotional and practical terms. Freedom and friendship are closely connected in Pahl's analysis:

> Friendship is a relationship built upon the whole person and aims at a psychological intimacy, which makes it, in practice, a rare phenomenon in its purity, even though it may be more widely desired. It is a relationship based on freedom and is, at the same time, a guarantor of freedom. A society in which this type of relationship is growing and flourishing is qualitatively different from a society based on the culturally reinforced norms of kinship and institutional roles and behaviour (Pahl, 1998: 111).

Pahl may be right in suggesting that identities are now constructed on a broader framework than that associated with kinship and the family and it seems undoubtedly true that people are freer to choose their associates. However, his view that modern friendship is a qualitatively different type of relationship to those bonds based on kinship or locality raises important questions about what Pahl understands by friendship and community.

Pahl draws on Aristotle's distinction between friendship based on utility, on pleasure and on character to suggest that there are three types of friendship found in modern society:

1. Acquaintanceship,
2. Family-like friendships and
3. Pure friendship.

However, Pahl has little to say about acquaintanceship or family-like friendship and does not appear to regard these types of friendship as particularly significant for personal or social development. He argues that 'much friendship is little more than acquaintanceship' which he describes as 'a category of friend that we are happy to compartmentalize in different sections of our lives ... such social friends ... are useful friends and we generally keep a close watch on the balance of reciprocities' (Pahl, 1998: 104). Family-like friendships seek to reproduce for adults what an ideal family would be like for children and he gives examples of a supportive mother figure or a mentor relationship. While he argues that 'such friendships provide help, comfort and protection from the impersonal and sometimes harsh and competitive world of employment and market relations', Pahl also suggests that 'people who need these kinds of friends are still trapped in family-like roles and may not have the confidence to move beyond them' (Pahl, 1998: 104).

Pahl regards 'pure friendship' as the basis for a new style of social relationships: 'Arguably, the modern, pure form of communicative friendship is the most personal relationship possible, expressing an ideal of voluntary personal agency' (Pahl, 1998: 107). Pahl and Spencer refer to a 'true soul mate who help us to become the sort of person that we are, who form and affirm our identities and who may serve as one of the essential threads that bind our biographies together' (Pahl and Spencer, 1997: 104). Much of Pahl's argument about modern friendship is premised on a belief that family relations inhibit personal growth and development and prevent a person becoming their 'true self': 'Increasingly, people are seeking to escape the sometimes suffocating, if not damaging and emotionally disturbing, aspects of their family of origin.... Adult relationships free from dependency and the emotionally suffocating echoes of family life reflects an individual's growing maturity' (Pahl, 1998: 106). Drawing on Aristotle's concept of 'character friendship' Pahl suggests that true friendship is about 'knowing and being known' and 'being seen in the round' (Pahl, 1998: 105). Yet Pahl argues that friends may not wish to know anything about the friends or families of their friends: 'We are not obliged and we may not wish to be the friends of the friends of our friends or to know anything about their families. They come as they are' (Pahl, 1998: 105–106). If true friendship is about knowing another person, it portrays an extraordinarily limited concept of authentic and meaningful relationships if important defining elements of a person's identity such as one's family are of no interest or importance in getting to know that other person.

Pahl's understanding of friendship is surprisingly decontextualised. This seems to be a reflection of his view of community as representing coercion and constraint and of modern friendship as representing choice and individuality. While Pahl does acknowledge the importance of trust as a precondition of friendship, his understanding of mutual trust seems somewhat limited to a type of trustworthiness within a dyadic relationship and to ignore the social context within which trusting relationships develop, for example the neighbourhood, club, church, pub and other intermediary groups and institutions.

Furthermore, Pahl sees 'true' friendship in rather narrow individualistic terms that precludes the possibilities of friendships developing among people with whom we are involuntarily tied such as neighbours and kin. He also misses the wider social significance of friendship which we argue can help to create worlds of meaning, promote practical social concern, engender a sense of belonging and even on occasion lead to positive collective action. Therefore, we would argue that choice and intimacy are too restrictive as conditions upon which friendship in all its guises and aspects can flourish.

Pahl's restrictive understanding of both friendship and community is also seen in his use of the concept of a 'social convoy' which he explains in terms of 'a changing cluster of significant others who provide affective, relational and normative support as we move through the life course' (Pahl and Spencer, 1997: 104). Pahl appears to view community as little more than a changing cohort of significant others thereby presenting a view of the individual as autonomous and free to pick and choose among potential relationships:

> The social convoy that we associate with as we go through life – our fluctuating and ever-changing personal community – defines us, reflects us, supports us and so much else besides (Pahl, 2000: 167).

In viewing society as 'micro-social worlds based on achieved friend-like relationships' Pahl sees individuals as separate, autonomous and making connections for very instrumental purposes. There is no understanding of relationality itself or of the importance of intermediate-level groups in creating connectedness. While he claims that encouraging 'friend-like relationships of choice' can help to develop 'the stronger moral community' and that there is 'no point in relying on the confused rhetoric of community', Pahl's understanding and vision of the moral community is surprisingly thin and lacks understanding of the transformative quality of friendship.

Graham Allan in his review of kinship and friendship in contemporary society gives a more nuanced account of what communal relations entail and what public and private sociability actually is. Unlike Pahl, he does not see friendship as belonging only to the private sphere arguing that: 'Friendships are not just freely chosen. They are developed and sustained within the larger framework of people's lives' (Allan, 1996: 100). Allan therefore holds out the possibility that friendship can tell us much about the relationship between the individual and the social – the nature of sociality. He discusses certain key characteristics of friendship such as the idea of friends being equal and of exchanges balancing out over time (Allan, 1996: 89), and the tendency for friends to have roughly the same characteristics as oneself (Allan, 1996: 91–92). Class and gender are thought to be important determinants of structural location that influence friendship patterns through the development of certain interests and the availability of material resources.

Allan argues that what is important is not solely to do with 'best' or 'closest' friends but rather with 'how the range of different relationships in an individual's "sociability network" is used within social life' and he goes on to raise questions about 'how friendship contributes to social life, and about the ways in which friendship ties in with other aspects of social organization' (Allan, 1996: 107). While Allan raises these important questions about the density of sets of relationships with friends and kin and also about the sense of obligation and commitment we feel towards particular ties, his overall conclusion is that personal friendship networks are largely unique to each individual (Allan, 1996: 127–130). However, we would argue that a limitation of Allan's approach is that it can miss one important form of sociality, that of generalised friendship, by concentrating on individual-centred friendships.

Pat O'Connor also discusses the limitations of focusing exclusively on intimate friendships and seeing them as the epitome of friendship. Thus she highlights other faces of friendship which she argues exist in the public arena and which are deeply embedded in social structures. She describes friendships among men associated with traditional male pub culture in terms of a kind of 'inarticulate solidarity' which is reflected in routinised male interaction. She describes similar relationships among women associated with group activities such as singing and telling stories in large family-like groups (rather than intimate confiding in dyadic relationships) which imply ties of sociability (O'Connor, 1998: 125). She also points out that relationships where emotional concern and practical tending are offered are often perceived as 'quasi-kin' friendships (O'Connor, 1998: 125).

O'Connor points out that in a variety of different cultures friendship as a cultural form transcended the private world entirely. In Ireland in the1930s as in other European countries relatives were described as friends (Arensberg and Kimball, 1968). Hannan noted that the concept of friend as a freely chosen confidante was still used by older people in rural areas to refer to kin (Hannan, 1972: 176).

In her review of the literature on friendships among older women O'Connor draws attention to what she sees as undue emphasis on the psychological conditions of attachment and the effects of friendships on emotional well-being in general and in specific situations of difficulty or transition (O'Connor, 1992: 119). She argues that sociability, rather than the intimacy that is often narrowly defined as confiding behaviour, is the key element in many friendships (O'Connor, 1992: 153). She highlights the complexity of the concept of friendship in relation to older women giving examples of different types of friendships:

> It is obvious, then, that the question 'Do the old have friends?' is a deceptively simple one. The answer to it presupposes a degree of consensus about the definition of friendship which exists neither amongst researchers nor amongst those interviewed. Variation in the trends emerging in these studies partly reflects these various definitions and partly variation in the nature of the samples used. It is possible to identify the various types of friends implicit in these studies: for example, the purely sociable companion: the long-standing but not interactionally active relationship; the very close intimate and/or identity-enhancing relationship; and the pseudo-kin friendship (O'Connor, 1992: 124–125).

O'Connor argues that the most supportive relationships may be those which provide support, without it being asked for, as part of the interactional dynamics of the relationship; and/or support which is offered in such a subtle way that no-one loses face or has to admit their need for help (O'Connor, 1992: 30). While she appears to express friendship in more political than explicitly moral terms, she too identifies the important element of obligation in understanding friendship. Her criticism of much contemporary work on friendship among older people is that it has remained implicitly within a functionalist tradition. She argues that little analytical attention has been paid to the fact that the social and cultural position of older people – especially poor older people – [and perhaps disabled older people] makes it difficult for them to have the resources necessary to create and/or sustain relationships. In reviewing work that examines the effects of marital status, life stage and class position on women's friendships, O'Connor suggests that it may be only in old age that structural and situational factors (for example reduced caring responsibilities, economic independence, and particular public interaction venues) can be seen as fostering the development of same-sex friendships. She also argues that the consequences of friendship for the societal structure have been little explored and suggests that sociable relationships between women, preferably at a group level, seem to have the potential, in the long run, for bringing about social and cultural change (O'Connor, 1992: 188).

Friendship and social participation in Rathmore and Rathbeg

In terms of social participation friendship appeared to be a major aspect of the intrinsic satisfaction that motivated individuals to engage in particular types of social interaction. We found many different types of friendship among older adults in Rathmore and Rathbeg. One-to-one friendships could be based on sharing life space or life experiences, for example shared experiences of living in a particular setting such as a sheltered housing complex, or rearing a family in a particular neighbourhood, or friendship arising from common interests involving a social club or hobby. Friendships could be long-standing relationships from childhood, or special relationships with kin who lived in the locality. The form the friendships took varied and included supportive friendships that helped people to 'get by', friendship among neighbours that was often expressed in friendliness and practical acts of kindness, friendship among members of clubs and social groups that was characterised by sociability and fun and generalised friendliness that involved acknowledging, greeting and being interested in the lives of others in one's locality.

Friendship in clubs

Friendship and friendliness pervaded the social groups in Rathmore and Rathbeg. A sociable type of friendship associated with shared leisure activities was observed in clubs and pubs. Friendship was especially evident in social groups where older women predominated. We discussed in Chapter Four the significance of participating in groups such as ARAs, Ladies' clubs and Senior Citizens' clubs. We observed friendships in the easy and warm conversations that seemed a spontaneous part of such social gatherings. Members could be observed inquiring about each other's families, exchanging news of mutual acquaintances and showing concern about each other's well-being. At a visit to a Senior Citizens' club the author noted the warmth of the greetings exchanged between one of the committee members and a number of the participants.[1] The organisers of the social activities at the club were seen to affirm each other's work; in one case a committee member praised another member because the club participants had said that she was someone to whom they could talk about their personal worries and problems. Another committee member said that as a committee they all got on well and remarked that 'there is no competition between us'. O'Connor's contention that sociability, rather than intimacy is the key element in many friendships appeared to be borne out by our data.

Jerrome's anthropological study of sociability in old age demonstrates the significance of social clubs and church based groups in the lives of older people (Jerrome, 1992). Jerrome suggests that joiners are distinguished from non-joiners in several respects: they are socially connected; they have high morale and they share a model of ageing which stresses struggle and resistance (Jerrome, 1992: 189). She argued that participating in such groups provides enormous benefits including companionship, sense of worth and sense of identity that can transcend differences and create connectedness:

> The pattern of multiple membership produces social networks which are both extensive and close-knit. .... The set of shared understandings and values evident in the culture of old-age organisations binds participants together as members of a moral community which transcends differences of gender and social class (Jerrome, 1992: 189).

In exploring the meaning of participation, Jerrome highlighted both common and different orientations among participants. She found sub-

1 Visit to a bingo session of a day centre in Rathmore on 10/11/03.

stantial agreement about things such as the importance of friendship, being active, what was appropriate behaviour for peers and about the value of tradition. However, personal motivation could range from passing the time to having an opportunity to help others or even evangelising activity (Jerrome, 1992: 193).

While Jerrome saw participation primarily in terms of a personal strategy for successful ageing, she introduced an interesting concept, that of 'social integrity' to explain the meaning behind participation:

> Whether in the church or outside it, old people in organisations – perhaps all old people – are engaged in a search for social integrity. They aspire to behave in ways that which are consistent with their best images of themselves. ... In the act of attaining it [social integrity] they find other things too: companionship, interest, a good time in the company of peers. ... Friendship is a major resource and an aid to survival (Jerrome, 1992: 195).

The value of Jerrome's ethnographic study is that as well as describing in rich detail the lives and concerns of older people, she explores the symbolic significance of their interactions for maintaining a sense of purpose and identity. She highlights in particular the importance of friendship in trying to make sense of the existential problems of life.

Generalised friendship

We have already spoken about how much our respondents valued their good neighbours. We described the helping tasks that neighbours did for each other and we referred to the many spontaneous comments made about the kindness, helpfulness and dependability of neighbours. However, friendliness was not confined to neighbours who the respondents knew well and who they classed as friends. We found that the practice of meeting and greeting people was widespread. Respondents were asked whether people greeted them when they go out in the street and whether they regarded this as important (Q28a and 28b). In Rathmore all but three people replied that people did greet them. Over

80 per cent of respondents said it was very important to be greeted, 14.5 per cent said it was fairly important while only 3 per cent said it was not important. In the more rural setting of Rathbeg it was not surprising that all eight respondents said that people greeted each other and all agreed that this was very important.

Having conversation with people on the way to or from church or to and from the shops was important for many people. For example one female respondent in Rathmore said that she would meet neighbours up at the shops and that she would 'talk to everybody'. We were told by some male respondents about conversations they would have with other men while watching football matches or training sessions in the local park or on the way to or from Mass. These conversations sometimes involved parties who did not know each other's name, and while this form of social intercourse might be assumed to mark a person as a 'mere' acquaintance rather than a friend, this did not appear to detract from the importance of that habitual practice of social intercourse for the sense of affirmation it provided. Another indicator of generalised friendship was giving support to local fund raising activities and attending funerals of neighbours.

Edmondson, in describing these communicative forms which she found in West of Ireland culture, argues that this form of 'distributed intimacy' can be contrasted with more concentrated forms of intimacy found in industrialised societies (Edmondson, 2001: 68). Edmondson suggests that emotional satisfaction may be derived from daily contacts spread throughout the locality in which one lives and not necessarily resting only on intimate and personal relationships.

The importance of obligation as an inherent element of friendship was a theme that emerged very strongly in our research. When describing helpfulness among neighbours in her locality in Rathmore one lady spoke about an obligation to be kind to people: 'If something is wrong everybody gets together and does something. No matter what people's faults are you have to be kind to them.' Another respondent in Rathmore who was not involved in many formal groups demonstrated a strong sense of empathy and obligation to others in her account of why she greeted passers-by when she was out walking:

> It's very important to make eye contact...it's especially important to say hello to older people...you might be the only person to speak to them that day ... that's what community is all about.

This sense of openness to other people was also evident in Rathbeg where people routinely greet each other on the street, at church, in the pub and in the shops. Generalised friendship is observed in the sympathetic way that people inquire about other people and speak about other people in the community who may be experiencing difficulties. It is also seen in practices such as visiting people in hospital whom the visitor might not visit in their homes and in practices associated with funerals involving paying respects to the deceased person and visiting bereaved families. It does not mean that all residents are friends in the sense of spending time enjoying one another's company or sharing intimate knowledge about one another's lives. However, it does indicate a form of friendship that involves people co-creating a world that makes lives meaningful and produces tolerance of others. Practices involving generalised friendship tend to be overlooked in accounts of friendship that emphasise only close supportive personal relationships.

Friendship: class and gender differences

Social class and patterns of sociability

In Chapter Four we discussed differences between working class patterns of sociability and those of the middle classes. The literature has suggested that traditional male working class patterns of socialising with non-kin tended to be confined to specific settings whereas for the middle classes there was a 'cross-over' between different foci of activity. The argument is that this permitted working class men a greater degree of control over resources for people with limited means and allowed them to maintain their social standing and to manage reciprocities (Allan, 1998: 79). More recent theories of privatisation have suggested that the idea of the home as an arena in which men can find expression and fulfillment changed patterns of sociability and styles of friendship (Allan, 1998: 83). However, Allen points out that it is overly simplistic to depict male working class sociability as being either 'communal' or 'privatised' (Allan, 1998: 84). Allen points to evidence of changes in sociable practices which suggest that non-kin sociable relationships are more important than privatisation theories had suggested and that there may be a greater similarity between the classes in patterns of sociability than earlier studies found (Allan, 1998: 86, 87).

Nonetheless, the ability to pursue certain interests and the degree to which networks of social relations are locality based is often a function of economic constraint. Having a car to travel to an activity or the money to join a golf club for example may differentiate older people in terms of their lifestyle. Furthermore, different opportunities associated with class influence the extent to which people have educational and cultural experiences that develop their interests and creativity. We found lower levels of social integration in a more socially disadvantaged part of Rathmore that impacted on older people's opportunities to become involved in enjoyable social activities. While not statistically significant, we found that respondents who belonged to the middle classes and respondents who had higher levels of education were more likely to be found in the more connected group. We noted Hall's contention that the differences in levels of social capital available to the middle classes and the least privileged had in fact widened in contemporary affluent societies.

At the same time class differences were not greatly evident in many aspects of sociability. We found that participation in outdoor sport was not strongly related to social class while participation in Active Retirement groups was fairly even across white-collar and manual groups. Many of the localities in Rathmore could be described as socially mixed and clubs and pubs are venues where people from different socio-economic backgrounds go. What may be more significant are cultural perceptions and practices associated with a form of enforced communality in areas that are more homogeneous in terms of social mix. These cultural practices may hinder the development of more purposive or educational forms of leisure and social activities in some areas. However, by making creative leisure opportunities available, older people, even if they never pursued such activities before, have the possibility to experience new learning, to develop themselves and to enjoy sociable companionship.

Gender and patterns of sociability

Differences in men's and women's friendships observed in the study were associated with gender differentiated roles and patterns of habitual interactions. For example sociability and support characterised the friendships we observed among women in settings such as Senior Citizens' clubs and Ladies' clubs and in one-to-one friendships among women in sheltered housing. On the other hand sports clubs and pubs were settings where men enjoyed social interaction. Men's friendships were formed through competitive sport, special interest and activity focused groups as well as in settings such as pubs. While the form of sociability associated with male pub culture has been described as 'inarticulate solidarity' and may be dismissed as less significant in terms of connectedness than more affective female friendships, it has been contended that the differences between women's and men's friendships diminish markedly as the strength and duration of the friendship increases (Wright, 1982). In a visit to a pub in Rathmore our male informant was able to tell us about the people who were present and the wide range of topics that were discussed. While men may eschew the 'chat and crafts' image of the more traditional type of social group for older people, Active Retirement groups with their emphasis on activities such as indoor sport have great appeal for men. However, we also found that ARAs could offer an environment for women who did not want instrumental and task-oriented activities but rather conversation and more passive participation.

It is important to note that while some men may have disliked the ethos of certain older people's groups and attached greater importance to individual autonomy, conveyed in such remarks as 'We were reared to be independent in my day', we found no difference between men and women in their overall levels of connectedness. The study underlined the extent to which men contributed just as significantly as women to the lives of neighbours and others in their localities through practical helping tasks such as gardening and house maintenance tasks and through voluntary work such as community enterprise projects, charitable work, Credit Union and Residents' Associations.

The study has drawn attention to the importance of friendship and the value of clubs and groups in facilitating the possibility of satisfying and sociable peer relationships. At the same time we have described people who had few friendships, little variety of social interaction and who experienced loneliness. Concern has been expressed about some older men in particular who are not attracted to clubs and day centres and who appear to lack companionship and support from peers. In order to understand better why older men are not attracted to traditionally run organisations and clubs for older people Davidson *et al.* undertook a research project that examined older men's involvement in different types of formal social organisations in South-East England (Davidson *et al.*, 2003). The authors found that the older men they interviewed had usually belonged to the same organisation for many years. Furthermore, most of these organisations provided opportunities for both social interaction and for active pursuits such as sports and/or 'useful' activities such as carrying out voluntary work (Davidson *et al.*, 2003: 84). In contrast to clubs such as Lions and Rotary, clubs geared specifically for older people such as day centres were viewed by older men as providers of passive pursuits and there was a stigma attached to them (Davidson *et al.*, 2003: 85). Their findings suggest that having opportunities to contribute positively to the lives of other people is important for older men:

> We suggest that the importance of activity and 'doing something useful' does not diminish with advancing age as long as the older man maintains reasonable health and mobility ... Clubs which catered for 'women and old people' were viewed as a last resort or, more commonly, a place they 'wouldn't be seen dead in'. However, there are lessons to be learned from the club we studied where older men attended and considered it to be a 'normal' organisation (Davidson *et al.*, 2003: 88).

The authors' conclusions have important implications for policy makers and service providers of community supports for older people. It has been suggested that the ethos of many day centres appeals to women and not to men and there have been calls in recent years for activities to be more 'men friendly'.

Friendship and moral frameworks

A sense of openness to the lives of others and of social concern was evident in many of the social groups we observed. We noted that some groups in particular had good inclusive practices, for example we have already discussed the ARA in Rathmore where people were encouraged to take part at whatever level of activity they were comfortable with. We came across many examples of older people giving encouragement to each other in unobtrusive ways that demonstrated kindness and understanding. It was reported that neighbours and club members in Rathmore would look out for someone who was developing symptoms of dementia. This would involve continuing to include them in their own daily routines such as asking them to walk to the shops with them, encouraging them to participate in local social activities and being attentive to their safety when they were out and about.

Furthermore, the idea of 'good example' was suggested by the lives of many individuals who were active in groups or exemplars of positive neighbouring. The actions of such people create expectations and set standards of what can be aspired to in terms of social and cultural integration. For social groupings to produce beneficial social outcomes a shared understanding of the good of the group is vital. The friendships found among social groups in this study appeared similar to a form of civic friendship described by Bellah *et al.* in *Habits of the Heart* in which the authors extolled traditional ideas of civic commitment in American society and expressed concern that this was being undermined by the culture and language of individualism:

> Traditionally it was the virtues indelibly associated with friendship that were central to the 'habits of the heart'. It is also part of the traditional view that friendship and its virtues are not merely private: they are public, even political,

> for a civic order, a 'city', is above all a network of friends (Bellah *et al.*, 1996: 116).

Putnam has emphasised the interactive effect between friendship, reciprocities and good example in his discussion of the ripple effects of doing good:

> Volunteering fosters more volunteering, in both formal and informal settings … studies have shown that (with other social and personality traits held constant) people who have received help are themselves more likely to help others, so that simple acts of kindness have a ripple effect. In short, giving, volunteering, and joining are mutually reinforcing and habit-forming – as Tocqueville put it, 'the habits of the heart' (Putnam, 2000:121–122).

We have observed that many individuals who were community oriented and/or kind and helpful to others often had themselves experienced personal difficulties and had gained personal strength from helping others. We also observed that people recognised good work, valued it and were inspired by it or influenced to see the world as a more habitable place. Thus, the types of friendships evident in some of the formal networks of older people, especially those with a self-help or community development focus, appeared to offer potential for promoting both a satisfying old age and raising cultural awareness and sensitivity to the 'the other' in our communities.

Friendship, moral frameworks and community

Bellah *et al.* refer to the writings of Aristotle in highlighting three essential components of friendship: friends must enjoy one another's company, they must be useful to one another, and they must share a common commitment to the good (Bellah *et al.*, 1996: 115). If Aristotle was right in claiming that one of the main duties of a friend is to help the other to be a better person it is difficult to see how friendship can be understood without reference to community. Bellah *et al* argue that individuals need the nurture of groups that carry a moral tradition reinforcing their own aspirations (Bellah *et al.*, 1996: 286). They suggest that what differentiates any group from a community is a shared concern for the common good which is characteristic of community (Bellah *et al.*, 1998:17).

Charles Taylor proposes that one of the most basic aspirations of human beings is the need to be connected to what is of fundamental value or what an individual sees as good (Taylor, 1994). Taylor argues that it is a condition of a functioning self to speak from a moral orientation and to this end we need frameworks that provide goals outside our own desires and inclinations. These frameworks enable us to make qualitative distinctions that define our direction in life:

> Because we cannot but orient ourselves to the good, and thus determine our place relative to it and hence determine the direction of our lives, we must inescapably understand our lives in narrative form, as a quest (Taylor, 1994: 51).

One's moral and spiritual orientations are key elements in self-definition along with family, social space, social status, function and one's intimate relations:

> I define who I am by defining where I speak from in the family tree, in social space, in the geography of social statuses, and functions, in my intimate relations to the ones I love, and also crucially in the space of moral and spiritual orientation within which my most important and defining relations are lived out (Taylor, 1994: 35).

Taylor argues that one's sense of self involves one's stand on moral and spiritual matters as well as some reference to a defining community (Taylor, 1994: 32). A vision of the good becomes available for the people of a given culture through being given expression in some form, otherwise certain goals and values are not even possibilities – 'articulation is a condition of adhesion'. According to Taylor a sense of the good finds expression in linguistic descriptions, prayer, ritual and visual symbols (Taylor, 1994: 91). Therefore, Taylor argues, the good needs to be articulated so that people can have available to them possibilities for making qualitative distinctions or moral choices.

We would argue that friendships in Rathmore and Rathbeg can be seen as involving moral orientations and processes through which people try to orient themselves to 'the good'. The concept of friendship may therefore be the key to understanding how social networks and norms of trust and reciprocity in socio-spatial settings can become translated into a good life and good communities.

Conceptualising friendship in Rathmore and Rathbeg

Our view of human development and sociality sees individuals as co-creators of social worlds that give meaning to life. Older people in Rathmore and Rathbeg create worlds of tolerance and respect through friendships of different kinds. Friendship has the potential to create more habitable social worlds. The lives of many older people embodied a strong sense of solidarity with others through both particular and generalised friendships. We argue that showing practical concern for non-kin through informal helping and/or voluntary work epitomises forms of friendship and social concern that constitute both an important social resource and that appear to indicate a special kind of connectedness. A respondent in Rathmore, who had been involved in many voluntary and community initiatives over the years, described herself as: 'very community based, knowing what's going on and sharing in joyful and difficult events.' A respondent in Rathbeg who was involved in a wide network of social groups, informal friendships and voluntary activity described her major interest as 'social awareness and meeting people'. In both cases friendship, positive neighbouring, social engagement and voluntary work were intertwined in their lives. At the level of sociability taking part in enjoyable shared social and leisure activities promotes mutual understanding.

Pat O'Connor's work on friendship among women has helped us to understand the importance of different types of friendship among older people and to appreciate the social and cultural conditions that facilitate or inhibit the emergence of particular types of friendship. O'Connor identifies sociability and affirmation as key elements of friendship. We have noted manifestations of friendship in significant and subtle forms of caring behaviour especially directed towards others experiencing some personal difficulty. We encountered close friendships among relatives and we found much evidence that relationships among neighbours could develop into friendship.

The socio-cultural context is important in understanding the opportunities for women and men to develop friendships at different stages of the life course. We encountered many instances where women got involved in community and voluntary activities when their families were reared or following widowhood and where they experienced friendships that enhanced their lives. Yet, some of these women said they would have liked to have had the opportunity to become involved earlier in their lives and implied that this was not possible. Anecdotally we were also told about older women who lead housebound lives and who suffered from depression. Due perhaps in part to the constraints of a narrowly defined gender and class position, these women may not have had opportunities to develop a strong sense of self in their older years.

We have argued that when investigating friendship it is important not to rely exclusively on interview data since this may omit cultural practices such as greeting people which respondents might not actually verbalise. We have provided evidence of a form of generalised friendship involving greeting and chatting to people who were not close friends. Such forms of sociality are often dismissed as involving weak ties and routinised interactions. However, we would argue that such interactions can be highly significant. Granovetter argued that weak ties are 'indispensable to individuals' opportunities and to their overall integration into communities' (Granovetter, 1973: 1378). Granovetter's primary interest was in the extent to which weak ties gave access to useful social knowledge such as information on job seeking. However, this study highlights how so-called 'weak ties' involved in acknowledging others with whom one is not close in one's locality can in fact represent important forms of connectedness by affirming someone's 'social integrity' and helping to create a sense of belonging.

Summary

We have argued that some contemporary accounts of friendship fail to capture the multidimensional aspects of the concept. Friendship is a relational concept characterised not only by intimacy but also by solidarity, sociability and respect. We encountered friendships in a variety of ordinary settings in Rathmore and Rathbeg and by no means confined to people's private lives. We argued that writers such as Raymond Pahl have overstated the importance of psychological closeness and choice as defining features of friendship and have given insufficient attention to both sense of obligation and development of shared moral understandings among friends. We have also demonstrated the extent to which place based friendships involve moral orientations and processes through which people try to orient themselves to 'the good'. The present study found strong ties among neighbours that did not necessarily involve intimacy or sharing of confidences but rested on respect, common values and shared experiences. Neighbours who had been involuntarily placed together had developed close friendships over many years. We also found that in some cases kinship ties involved strong friendships that were a source of both sociability and support. We noted in Chapter Seven that in Rathbeg cousins were often regarded as friends and confidants.

We discussed the institutional settings and intermediate groups that provided significant social contexts within which trusting relationships developed and friendships were fostered. Friendship appeared to be a key aspect of the intrinsic satisfaction that individuals derived from social engagement. While it is important to take account of the effects of gender, class, marital status and life stage on the opportunities for and possibilities of forming friendships, we found that friendships of different kinds flourished in ordinary everyday interactions in neighbourhoods and particularly in social, leisure and voluntary groups. Our findings would very much concur with Jerrome's that social clubs provide enormous benefits in terms of companionship and self affirmation. However, there are clear public policy implications that arise from this which we will discuss in the final chapter.

Having earlier identified barriers to social participation including socio-economic constraints, cultural attitudes, institutional norms and personal difficulties, we argue that public policy must seek to support the provision of sociable and purposive leisure opportunities at local level.

We identified generalised friendship in habitual cultural practices such as inquiring about others and greeting others in one's locality. This sense of openness to other people was evident in both Rathmore and Rathbeg and can be seen to represent a collective wisdom – a cultural resource that entails both personal and communal benefits. We argue that friendship is a key element in understanding the ways in which different types of connectedness translate into good communities. Friendship networks in both formal and informal groupings help to create social worlds that make life both tolerable and enjoyable. Thus the transformative potential of friendship for personal and social well-being has been demonstrated. Friendship has also wider social significance in helping to create worlds of meaning and in promoting practical social concern.

Chapter Ten
Ageing and Community in Rathmore and Rathbeg: Summary and Conclusions

Overview

The study has drawn attention to the often dense network of interaction among older people, and has probed the kinds of activity and relationships that make it up. To understand how older people live in their communities it is necessary to explore and portray a relational landscape which can sometimes seem in danger of being ignored in Ireland at the start of the 21st century. This relational landscape is rooted in geographical place; within it, the special contributions made by particular individuals and groups are recognised and respected. Thus, we are able to point to realms of activity which offer genuine meaning and value in older people's lives. Theoretical understandings of community and friendship highlight the importance of socio-spatial relationships for meaning in later life. Policy implications relate to the positive contribution of older people themselves to social and public life and the importance of recognising and fostering solidaristic relationships within localities.

To discover this we investigated the social and community participation of older people in an urban and a rural setting in Ireland. The aim was to examine relationships, commitments and social activities of older people that were associated with the neighbourhood in which they resided and with the wider community. The focus was on social and leisure activities, organised both formally and informally, volunteering, membership of clubs and groups, religious practices, informal helping activities and informal sociality. Two geographical areas were chosen for the study – Rathmore, a suburban area of Dublin and Rathbeg, a rural area in Donegal in the North West of Ireland. The study used mixed methods and adopted collaborative approaches to generating socially useful knowledge. By selecting two contrasting areas for ethnographic study and by making further selection of sites within those areas such as sheltered housing, clubs, special interest groups and pubs, we were able to get close to observing the networks of interaction among older people and to recognise the activities that were meaningful in older people's lives. In addition through creative design of a large-scale survey in Rathmore and a small survey in Rathbeg we investigated patterns of participation across relevant domains of communal life. Understandings of community and of what constitutes a good life in old age were key questions that were explored in the course of the study from the perspectives of older people themselves.

Participation in social activities in Rathmore and Rathbeg: leisure, clubs, religion and helping

Older people in Rathmore and Rathbeg engaged in a wide variety of both common pastimes and more individual social and leisure activities and participated in a plurality of groups. About half of the sample in Rathmore were involved in at least one group. The most predominant groups were church groups and sports clubs while community development groups, Active Retirement groups and charitable groups were also important. Communal leisure activities such as playing sport, going to the pub, attending a Senior Citizens' social club or special interest activities provided opportunities for intrinsic satisfaction and mutuality of experiences. However, ideas about enjoyable and worthwhile social activities were intertwined with ideas about voluntary work and community involvement. Deriving satisfaction from life was strongly related to opportunities for sociability, meaningful engagement with other people and developmental activities – activities that constitute what Erikson termed 'vital involvement'. Participation in groups, whether they were primarily leisure or more civic oriented, created connectedness and promoted both a sense of self and of community.

Older people experienced satisfying relationships through communal leisure activities that suited their own interests and preferences. The appeal of the more traditional model of Senior Citizens' club had waned somewhat while Active Retirement groups with a more participatory and activities focus appeared popular with a wider group of older people. Nonetheless, both types of group provided settings for enjoyable leisure and social activities, giving participants a sense of engagement with others and enabling them to take part at whatever level of involvement suited them. Even activities often thought to be passive and to involve little learning or creativity such as bingo or outings were found to promote connectedness in important ways. Special interest activities such as art and historical societies emphasised the creative and developmental aspect of life for any age group. These social and leisure groups were characteristically locally based, dependent on voluntary effort and were important settings where friendships flourished.

While public discourses on retirement and leisure tend to emphasise individualistic and consumerist values, we found that the social activities of older people involved sociability and friendliness and many activities were directed towards helping others. The personal and social significance of leisure activities suggests that, as Moody argued, leisure is much more than residual time away from more functional activities but is rather a multi-dimensional quality of life that relates to personal satisfaction and service to others.

Religious practice was strong and vital in the lives of the vast majority of older people in the study. Reflecting the religious composition of the population in Ireland, the overwhelming majority of respondents were members of the Catholic church. However, their adherence to institutional religious practice and the embeddedness of religion and spirituality in many aspects of their daily lives appeared contrary to secularising tends away from regular church going and towards more private spirituality among Irish Catholics. At the same time, religion still retains both personal and social meaning for significant numbers of the general population in Irish society. For the respondents church going and prayer were very important. While different approaches and emphases in relation to religion and spirituality were expressed by respondents, religion was vitally important in coping with life and as a source of ultimate meaning. Furthermore, the social ties embodied in Parish communities were an important element of connectedness in Rathmore and Rathbeg.

Older people contributed enormously to the functioning of their local churches and in church-based voluntary work. However, in the context of changes in institutional religious practice in Ireland, concerns were expressed by some respondents that the cohesiveness that church membership currently provides for older people may not be present in newer and future communities. It was often the case that the lives of those who contributed significantly to communal life through voluntary and community work were imbued with a strong sense of Christian duty or a disposition that was compatible with practical Christian concern for one's 'neighbour'.

Connectedness was also expressed and experienced through informal helping activities. Both intrinsic satisfaction and support derived from positive interactions involving children, grandchildren, other kin and friends. However, positive neighbouring emerged as a major social resource and was characterised by reciprocities, kindness, practical social concern and friendliness. People valued the kindness and helpfulness of neighbours that was demonstrated through many helping tasks, visiting and support at times of particular difficulties. Being attuned to the lives of others in one's locality was evident among volunteers and those involved in community activities and produced very beneficial social outcomes when voluntary work was harnessed to social policies, for example in the development of day centres and sheltered housing.

Solidarities, supports and changing cultural practices in Rathmore and Rathbeg

There were far more similarities than differences in the communal activities and significant networks of older residents of Rathmore and Rathbeg. Social networks were large, heterogeneous and dense in both localities. There was evidence of both local extended family and dispersed extended family in the two localities. Regular contact with and support from and to children was a strong feature of the lives of the respondents. This was evident in the frequency of visiting and telephone contact, help with aspects of daily living and enjoyment of family celebrations. While there was great variation in levels of contact with other kin, families of origin involving sibling support often became more salient as people aged. In Rathbeg where some older people had extensive local kinship networks, relationships with cousins were often significant for both sociability and support. Attachment to Rathbeg itself was linked with relationships and reciprocities built up over time and transcending generations. It should be pointed out that the author has strong family and cultural attachments to Rathbeg and, as a member of the Rathbeg diaspora, was better able to appreciate the details of attachment to Rathbeg than those of Rathmore.

The value attached to having good neighbours was a recurrent theme throughout the study and was a very important aspect of attachment to place. Positive relationships with neighbours were important both for people with extensive family, kin and friendship networks and for those with limited kin and friendship networks. Help given by neighbours to people experiencing illness and bereavement may be a stronger feature of Rathbeg than Rathmore. Nonetheless there were localities within Rathmore where cultural practices associated with traditional close-knit communities were observed, for example having Mass in individual houses or collecting money for a bereaved family. Indeed, the strong friendships that we observed among kin and among neighbours led us to question perceived boundaries between relative, friend and neighbour.

Another important form of connectedness was generalised friendship that involved acknowledging and greeting others who lived in one's locality. Having a chat with someone whom one did not necessarily know well generated a sense of solidarity and belonging. However, processes of social change in interaction with life course changes led to a perception among some older people that there was less contact among neighbours than in the past. Some suggested that since their children had grown up there was less occasion for interaction with neighbours.

There was a stronger sense of social/cultural change in Rathbeg than in Rathmore. This was reflected in respondents' spontaneous references to the changing composition of the community, changing social practices and different types of leisure activities which had led to some older people experiencing a diminution of contact and communication with others in their locality. While opportunities for intergenerational mixing outside family life were limited in both localities, there were more communal events that involved participation by older and younger people in Rathbeg than Rathmore. At the same time there was a perceived lack of community venues in Rathbeg for social and recreational activities for any age group. However, since a social centre was built there in 2006, which provides recreational activities and meals for older people on two days per week, there are now more opportunities for social interaction and developmental leisure activities. In Rathmore while respondents made reference to changes in their neighbourhood and the impact of growing older, this related more to life course changes than to socio-cultural changes per se.

One question that is difficult to answer is to what extent the forms of sociality traced in this study are specific to a country which until relatively recently had more semi-traditional forms of relationships than, say, England. While it is the case that many of the most significant relationships in Rathmore and Rathbeg depend on continuing residence over time, we also found many significant relationships that were not based on interactions over prolonged periods of time. For example, we found interactions and relationships among residents of sheltered housing in Rathmore that had considerable meaning and value in people's lives and that did not rest on long-term acquaintanceship.

Degrees of connectedness: differences related to class, gender and health

There was, not surprisingly, great diversity in both opportunities for positive engagement with others and the ways in which connectedness was expressed and experienced. Class and gender differences were evident in patterns of participation. Life chances and life choices are shaped and constrained by social class. While certain leisure interests were more likely to be favoured by white-collar respondents such as reading and crosswords, and activities such as bingo and going to the pub by respondents from the manual classes, other activities such as participating in sport or taking part in Active Retirement activities were not strongly differentiated by class. Class differences were strongly evident in membership of church groups and in volunteering. Middle class older people were far more likely to belong to church-based groups and more likely to volunteer than working class respondents. People from white-collar backgrounds are more likely to have the personal resources of education and know-how, confidence in dealing with bureaucrats and technical expertise to enable them to undertake projects and leadership roles in voluntary groups and clubs.

While much of Rathmore was socially mixed, there were areas that were more distinctly working class or middle class. In one particular locality in Rathmore there seemed to be a greater concentration of both personal and social problems, such as vandalism and overt drug taking, less positive sociality and fewer facilities for older people. This locality had a greater concentration of people from manual backgrounds many of whom came from a specific socio-geographic origin in the inner-city. A type of enforced communality may have limited the development of a wider range of educational and creative activities and led to reduced involvement in civic groups. In localities that were more socially mixed and with higher levels of income and education it was more likely to find a plurality and diversity of groups and better communal provision for older people. A combination of higher levels of stress (due to problems in families), poorer communal provision and environmental factors (such as vandalism) had contributed to lower levels of morale and a perception that older people had become somewhat isolated in this locality. A Senior Citizens' club had been recently disbanded and a number of older people said that this meant they had no social event to look forward to. Paradoxically, many older women in this area have taken on parenting responsibilities in relation to their grandchildren which was a far more burdensome task than the grandparenting experiences of their middle class counterparts, and apparently little acknowledged by health and social service agencies.

However, the effects of class on social participation appeared to be intertwined with cultural expectations and biographical influences. For example, while many of the residents of sheltered housing complexes belonged to manual backgrounds, the levels of positive sociality evident in some complexes seemed as much related to the vision and work of a number of residents whose cumulative life experiences involved many influences other than a simple social class classification. Influences other than social class ones included experiences of adversity and overcoming difficulties, experiences of support from others and religious beliefs. Participation in adult/community education, not necessarily related to social class, was another influence on positive sociality. Communal participation is influenced by a complex array of factors and is shaped by intersubjective understandings created through interpersonal relations in intermediate-level groups. Thus, within localities and social settings, individuals and groups co-create perceptions and expectations about what is interesting and worthwhile within contexts influenced by but not determined by the socio-economic backgrounds of the residents there.

Sociability and support were strongly evident in Senior Citizens' clubs and day centres that tended to be female-dominated. Active Retirement groups with their emphasis on activities such as bowls and art had greater appeal for men than more traditional groups for older people commonly characterised by images of 'chatting and crafts'. However, women tended to participate in greater numbers in the Active Retirement groups that we investigated. Policy makers have been concerned about the reluctance of men when they experience reductions in mobility to avail of facilities in clubs and day centres. It has been suggested that men perceive that the type of pursuits in such centres are passive and stigmatising and they would prefer to be doing something useful. It has been argued that insufficient account has been taken of the social worlds of men in designing day centres and community supports for older people.

The study demonstrated the extent to which sport was a major interest for men and provided some insight into men's friendships focused on competitive sport, special interest activities and forms of sociability and solidarity associated with regular pub going. The study also underlined the extent to which men contributed just as significantly as women to the lives of others in their localities through practical helping tasks such as gardening and house maintenance tasks and through voluntary work such as community enterprise projects, charitable work, Credit Union and Residents' Associations. We found no difference between men and women in their overall levels of connectedness. Therefore, any presumptions about women being more 'community oriented' than men were not borne out in the study.

While the study has drawn attention to meaningful activities and dense networks of interaction, many respondents experienced difficulties such as declining health, reduced mobility, death of spouse or friends, enduring or increasing caring responsibilities, insufficient health and social care supports and reduced income. Many experienced some degree of loneliness, and a small number were lonely often. Loneliness appeared to result from restrictions chronic illness placed on daily activities, lack of meaningful participation and loss of a close relationship. For example, several people spoke about how hearing loss prevented them from enjoying social interaction. Some people had very small kin networks, few friendships and had accumulated material and emotional disadvantages due to poor life chances. A number of people living in sheltered housing were described as 'reclusive, very negative, suspicious of each other and having become odd from living on their own', and residents and staff who worked to provide interesting programmes of social activities found it very difficult to get them to participate. Even older people who had children and who were visited regularly by them sometimes lacked companionship and opportunities for involvement with peers.

While there was a general perception that older people were well provided for by government with many people referring positively to free travel, free primary health care and increases in state pensions in recent years, there were specific areas of dissatisfaction including lack of a separate bedroom in sheltered housing units and shortcomings in the hospital services. Several people believed that frail older people living alone needed much more support such as better home help provision, help with practical household maintenance, more day centre places and befriending opportunities. Barriers and inhibitors to communal engagement identified by the respondents themselves included absence of facilities and organised social activities, for example a community hall or day care centres, poor public transport, shyness and reclusiveness. Several people believed that there was not enough contact between older people and younger people. Furthermore, a number of older people thought it was better for older people to mix with younger people than with their peers only.

Community-ness and connectedness

Five variables that indicated connectedness were being a member of a *club or group*, doing *voluntary work*, *helping someone* outside one's household, being involved in *visiting* at least once a week and having *regular enjoyable events* that one looked forward to. The significance of these indicators is that they capture key elements of sociability, solidarity and support – processes that are central to our understanding of community. These five indicators reflect 'community-ness' as Calhoun defined it – face-to-face interaction, commonality of purpose, familiarity and dependability. Furthermore they emphasise the contribution of older people themselves to communal life. While these indicators of 'community-ness' are not unique to older people in a locality, they have particular relevance to the lives of older adults since they stress reciprocities built up over time involving commitments, moral orientations, learning and creativity enriched by life experience. The relationships and interactions that constitute community are similar to the type of multidimensional relationships described by Tonnies in his seminal writings on community in the late nineteenth century. Community as understood in this sense refers to the possibility of people co-creating and experiencing authentic and multi-dimensional relationships with kin and non-kin that were affirming and dependable. Thus community-ness is strongly associated with satisfaction with life and with perceptions of what constitutes a good life.

The lives of older people who were strongly connected within their communities were characterised by dense multidimensional relationships. They had many and diverse interests and commitments that included family, hobbies and voluntary commitments and they experienced enjoyment and satisfaction with life. The strongly connected people were varied in terms of marital status, living arrangements and gender. The overwhelming majority said they enjoyed life very much. It appeared that engaging with others through social activities, common interests and helping were all intertwined and often bound within a strong Christian/humanitarian commitment. In addition, the support they gave others and their voluntary/community work were generally appreciated and valued. These people represent an important social resource because of their example and encouragement to others, because they articulate values of positive neighbouring and because of the practical collective outcomes they help to achieve.

The lives of older people who were less connected within their communities revealed diversity and complexity in relation to their experiences of sociability and solidarity. Some were socially inactive due to illness or caring commitments and a degree of boredom and loneliness was often associated with enforced inactivity. Many appreciated their good neighbours even if they had little regular social contact with them. Among the less connected group were people who were self-contained or entirely family oriented. Some were contented with their own company and had interests such as reading, following current affairs or domestic interests. Very few people were unconnected on all five dimensions and even those who were less connected were generally visited and helped by their families. However, a small number of separated males did not have regular social contact with their children or with all their children. Marital separation and/or lack of a steady work career placed some older men in a difficult financial situation and may have limited their opportunities to pursue hobbies or engage in social activities.

The use of latent class cluster analysis allowed us to measure connectedness among our respondents in a more aggregated way. The analysis suggested that we could use two models of connectedness: one based on the five indicators described above and the second model based on four indicators by omitting visiting as an indicator. The first model presented three groupings: a connected group, an unconnected group and a group who were connected through helping activities but not through social and leisure activities. The second model presented two groupings: a connected and an unconnected group. Factors associated with being connected (including the small group who were characterised by their helping activities rather than visiting or enjoyment of communal leisure) included subjective health, age, functional ability, having a major interest, whether or not respondents said there were others who relied on them and frequency of visits from relatives, neighbours and friends. Factors that were strongly associated with connectedness although not statistically significant were enjoyment of life, importance of religion and social class. Marital status, gender and being co-resident or living alone were not significantly related to connectedness.

The finding that visiting may or may not be included as an indicator of connectedness coupled with analysis of the significance of contact with relatives, friends and neighbours, highlighted the limitations of using visiting as an indicator of connectedness and the complexity of explaining the nuances of relationships involving relatives, friends and neighbours.

There are many ways in which one can participate in one's community as an older person that give meaning and purpose to life. It was not essential to participate in groups or clubs or to be physically robust and energetic to be involved in meaningful reciprocities with others in one's locality. The majority of older people in this study are neither strongly socially active nor socially inactive.

Ten relational patterns were identified based on relationships with significant others as well as communal participation through social and leisure activities, voluntary work and informal helping. The typology of connectedness takes account of the different ways in which older people contribute to the lives of others, whether family, other kin, friends or neighbours, and their social contacts and group affiliations across a wide range of arenas. For example, some people are involved in dense social networks across many types of relationships while others are more family centred or involved with a small number of friends. The typology also takes into consideration circumstances, transitions, opportunities (including those that arose and were created during the life course as well as existing opportunities in the social environment) and moral orientations and preferences. It is therefore a measure of both how 'vitally involved' our respondents are and their everyday interactions framed within the socio-cultural parameters of their lives.

Friendship

The concept of friendship emerged as a central analytical idea that helped to explain the types of experience that give meaning to older lives and highlighted the significance of relations in socio-spatial settings. We observed different types of friendship: one-to-one friendships, either long-standing or more recent, that were supportive and helped people to 'get by', friendship among neighbours, friendship among members of social groups that was characterised by sociability and fun, and friendship with kin. Furthermore, intermediate-level groups such as clubs, church, voluntary groups and pubs were settings where positive sociality was fostered and friendships developed. Satisfaction and genuine dependable relations were attained through the intersubjective experience of friendship among participants in communal social activities. Furthermore, place-based friendships helped to create enclaves that encouraged tolerance and other-regarding behaviour.

We do not mcan to suggest that only place-based friendships were satisfying or meaningful for older people in the study. Peer friendships flourished with those who shared similar interests outside one's locality. We have stressed the importance of satisfying leisure for developing and maintaining one's sense of self. Being able to participate with others in mutually satisfying leisure and other social activities may or may not be locally based. However, authentic relationships were fostered in many communal settings including those with a recreational purpose and those more purposively civic or community focused. Friendship appeared to be strongly contextualised in the daily interactions of older people in Rathmore and Rathbeg. In both rural and urban settings, interactions through communal leisure activities, helping activities, religious/voluntary groups and informal interactions helped to create social and moral worlds where people experience a sense of belonging, involvement and fulfilment. We also described a form of generalised friendship that involved acknowledging, greeting and being interested in the lives of others in one's locality.

The findings suggest the inadequacy of defining friendship primarily in terms of psychological closeness, choice and individuality. Rather they suggest developments of accounts of friendship given by Allan, O'Connor and Jerrome who highlighted the importance of sociability, affirmation and social integrity. The conceptualisation of friendship suggested here stresses its interactional nature and its transformative potential. The significance of the ordinary interactions and place-based relationships that we have described corresponds with the traditional idea of friendship as developed by Aristotle and Cicero and as understood in a number of Western religious traditions. We would argue that friendships in Rathmore and Rathbeg can be seen as involving moral orientations and processes through which people try to orient themselves to 'the good'. For example, we observed many instances of older people giving encouragement to each other in unobtrusive ways that demonstrated kindness and understanding. Thus, the social context in which dependable trusting relationships can develop is vitally important in the creation of a more humane environment.

We suggest that friendship among older people is best seen as a relational concept involving a number of processes such as fun, sociability, solidarity, obligation, respect and acknowledgement. We would also emphasise the tremendous value of clubs and groups of different types in facilitating the possibility of satisfying and sociable peer relationships. We argue that friendship is a key element in understanding what forms of social connectedness translate into good communities where older people are well integrated, live full lives and have opportunities to contribute to their communities.

Social policy implications

There are important implications for social policy that arise from our findings. If, as we have argued, a good life in old age must be understood in terms of community-ness, then older people, regardless of any limitations of physical or mental capacity, must have the opportunity to remain involved with others and engaged in meaningful activities and interactions. The evidence that older people contribute enormously to solidarities within their communities emphasises the need to recognise their contribution and the important social resource they represent. However, the extent to which older people are commonly perceived as unique individuals with a repository of life experiences and capable of continued learning and social influence in their neighbourhoods and society is open to question. Many studies and policy approaches focus on the older person as potentially lonely and as a recipient of services and supports. Fewer focus on older persons as givers of help and creators of social ties.

Thus, public policies must strive to nurture reciprocities and support and cultivate caring and friendship networks within localities. Studies have provided evidence that extensive state services for older people do not 'crowd' out family care but that formal and informal provision through state and family care are mutually supportive and lead to higher levels of help and support. This study portrays neighbourhood based caring and helping networks that give expression to a vast range of human needs. However, it is important that the informal caring work of families and volunteers is not taken for granted, and statutory bodies should attach priority to integrating their services with social networks within local communities. There were many examples of older people and others being involved in overtly constructing community through voluntary and community development work and we have documented the substantial gains that resulted from their work. However, we found that the extent to which public authorities and voluntary bodies work collectively and productively to develop care and social facilities for older people varies enormously from one health board area to another and indeed from one locality to another.

The idea of 'ageing in place' is increasingly being held up by policy makers as the goal of community care policies in Ireland. However, this implies that the surrounding community should be a good place to live in as an older person and this has clear implications for state provided community care services and for public policies in terms of creating a safe and congenial environment. In relation to day care for example it seems likely that the more locally based the centre, the greater the variety and opportunities for sociability and creativity that can be provided, and the more centres can tailor services to individual preferences, the more acceptable they will be to a greater proportion of older people. Therefore, an enabling public policy should focus on the following: pay greater attention to capacity building within the voluntary sector, provide more flexible and consistent support to families and carers and provide a wider and more integrated range of facilities for semi-independent living in localities.

The example of older people doing interesting and socially useful things and even the sight of older people in localities 'keeping going' can itself generate a sense of participation and meaning. Models of good practice in the provision of day care and sheltered housing encourage the participation of older people themselves in choosing and running social activities. In relation to care facilities for older people, both day and residential, links should be fostered and resources tapped within the surrounding community so that new opportunities are sought for creative development and the cultivation of new friendships and solidarities. The provision of social outlets including clubs, day centres and befriending schemes are particularly important for older people whose mobility is restricted or who may need encouragement to become involved in social activities. Public policies must be based on a view of all older people as having something to contribute to their communities. However, we need to give more thought to examining ways in which this could be done, for example, by intergenerational projects involving schools and exploring ways in which older people could help younger people.

The findings of this study provide tools for a type of 'community audit' that could be undertaken in localities and other social settings to evaluate the degree of 'community-ness' and the potential to develop public policy intervention strategies. The five indicators of connectedness – membership of clubs/groups, volunteering, helping someone outside the household, visiting and taking part in regular enjoyable events – could form a checklist against which community groups and public bodies could assess gaps in solidarities, supports and social ties, and evaluate the possibilities of activating the resources within intermediate groups and informal networks in neighbourhoods to generate more connectedness. It is argued that, for life to be satisfying for older people, they must be connected within enclaves of sociality that provide opportunities for sociability and friendship, for learning and personal development and for giving as well as receiving. Thus, personal and communal well-being for older people is facilitated by having a plurality of local groups and by strong positive neighbouring cemented by a supportive system of state welfare services.

Ageing, community and friendship

In conclusion we have found that older people in Rathmore and Rathbeg are involved in a dense network of interaction with family, kin, friends and neighbours. Positive neighbouring represents a particularly important social resource. Not least, friendships in places create connectedness; they have transformative effects on personal and communal well-being. Satisfaction with life and understandings of what constitutes a good life are strongly associated with meaningful engagement with others in one's locality. Through social and leisure activities, volunteering, informal helping, religious practices and informal sociality older people contribute enormously to communal life in ways that are often not recognised. Satisfaction with life would be enhanced for more older people and society would benefit from recognising and facilitating these positive interactions.

However, this represents a challenge for public policy and calls for critical reflection and debate on what is held to be of value in Irish society at the beginning of the 21st century. Ireland has become a wealthy and multicultural society in a relatively short time period. A glimpse into social changes occurring in Ireland can be seen in the concerns that grandparents expressed in a recent study about the influence of materialism, lack of discipline and sexual promiscuity on the lives of their grandchildren. Furthermore, religion in Ireland may not in the future provide the same basis for moral consensus and social solidarity as heretofore. In the light of such changes, this study could be a useful baseline against which future changes could be examined.

The study findings do not seek to evoke or recreate some 'imagined community' associated with a romanticised past. Rather, the study has documented and acknowledged authentic ways to live and identified the elements of a good life and good communities as understood by older people themselves. Celebrating, valuing and seeking to foster the activities that offer genuine meaning and value in the lives of older people would benefit not only older people themselves but their families, neighbourhoods and indeed society as a whole. Older people with their accumulated life experience and wise social practices are a vital resource in our society.

Appendices

Appendix A: Questionnaire – Participation of Older People in the Community

The survey is part of a PhD study on the social role of older people and their participation in their communities. We want to find out more about the positive contributions older people make in their communities and whether they have problems that are not properly understood.

What people tell us will be confidential and all participants will be anonymous – no names will be used in the survey results. We are most interested in an overall picture of life as experienced by older adults in a particular community.

To get a comprehensive picture of the experiences of older adults in Ireland today we would like to ask you questions on a range of topics. These will include your experiences of work and retirement, your leisure and interests, your family and social contacts, your educational experiences, the help you give and the help you receive in day to day living, your general state of health, your experiences of being an older person in this particular community and your experiences of using different health and social services.

We hope that the study will benefit older adults in the following ways:

- It will tell us about people's actual preferences rather than their perceived needs and this could help providers to focus services better.
- It will tell us about the diverse range of interests which older people have and this could help to suggest what social, recreational and educational activities might be supported in communities.

- It will show the various ways in which people help and support each other. This will tell us more about the quality of caring relationships in both the formal social services and informal helping relationships.
- It will tell us something about the quality of life of older people in communities and what factors affect this, such as health, income, local amenities and organised activities, earlier life experiences and so on.

It is hoped to hold a feed back session on the results later in the year to which all respondents will be invited.

Area code:
Interviewer's name:
Respondent's address:
Where interview took place (kitchen, sitting room etc.):
Date:

Instructions for Interviewers:

To be completed in respect of respondents aged 65 and over. Only one respondent to be selected per house. If respondent needs assistance in completing the interview someone from the household or a family member may assist. The respondent should be present and complete as much of the interview herself/himself.

1. Interviewers should ensure that an answer is recorded for each question by circling the appropriate code.
2. Interviewers should record any additional comments made by the respondents at the end of each page. ***Record direct words used by the respondents themselves as far as possible.***
3. When the interview is over check over the questionnaire and ensure that it is complete.
4. Complete parts B, C and D.

(Further instructions have been given on a separate sheet and in the training sessions.)

Part A – Interview Schedule

Present status and location

I'd like to ask you firstly about your own background and marital status, to help us build up a picture of the people who live in this area.

1. a) Could I just check what year you were born? 19_______
(must be before 1938)

b) Where were you born?

Elsewhere in Ireland (Outside Dublin)	1
Elsewhere in Dublin (Outside Dublin 5 area)	2
This part of Dublin (Dublin 5 area)	3
Outside Ireland (Specify______________)	4

2. Can you tell me about your present marital status and since when you have held this status.

	Status	Since
Married	1	19
Single, never married	2	
Widowed	3	19
Separated	4	19
Divorced	5	19
Married for second time	6	19

Work Background

I'd like to ask you about your work both current and past.

3. a) Which of the following best describes your employment situation at the moment?

Full time employed	1
Part-time employed	2
Self-employed	3
Unemployed seeking work	4
Unable to work because of illness/disability	5
Retired	6
Home duties	7

b) If employed or seeking work, in what occupation do you work / or do you intend to work?

Name of job ______________________________

Probe – ask about main tasks or duties if title of job is unclear

c) In what occupation have you (and your spouse/partner if applicable) spent most of your life since age 16? (Does not refer to full-time homemaker/ housewife)

	Self	Spouse/partner
Name of job:	________________	________________

Probe – ask about main tasks or duties if title of job is unclear

4. a) If retired from your main job, at what age did you (and your spouse) retire?

Self _____ Years Spouse _____ Years

b) For people (mainly women) who gave up work to raise family/engage in home duties – Did you do any paid work subsequently?

Yes ... 1 No ... 2

Name of job: ______________________

Length of time working in subsequent job/ ______________________

c) Were you in general happy to retire when you did?

Happy in general ... 1 Unhappy in general ... 2 Mixed feelings ... 3

5. a) Have you done any paid work since you were 65?

Yes, full-time ... 1 Yes, part-time ... 2 No ... 3

Record details

b) Have you done any paid employment in the past year?

Yes, full-time … 1 Yes, part-time … 2 No … 3

Record details

Pensions are an important source of income for many people and we want to gain an idea of how widespread different types of pensions are in Ireland today.

Are you in receipt of a pension? If 'yes' indicate type.

Own occupational	1
Spouse's occupational	2
Social welfare contributory	3
Social welfare non-contributory	4
Other (specify)	5
No	6

I'd like to ask you about any voluntary work you do and whether you are involved in any groups or organisations.

6. a) Do you do any voluntary work either in this community or elsewhere?

Yes … 1 No … 2

If 'yes' list up to three

1) ______________________________

2) ______________________________

3) ______________________________

b) Are you currently involved in any of the following organisations/groups? (***Go through each type***)

Specify a) level of involvement,
b) Organisation/activity and
c) previous if no present involvement.

Type	Occa-sional help	Regular involve-ment	Organisation /activity Specify name of and type of activity	Previous but not present involve-ment	No
Sports club (GAA, soccer etc.– training, managing, administrative etc.)	1	2		3	4
Charitable – (V. de P., Protestant Aid, Lions, social services work, meals etc.)	1	2		3	4
Parish/church based work (Parish office, newsletter, clean church, parish group, read at Mass etc.)	1	2		3	4
Community development work (resident's/ tenant's association, youth club etc.)	1	2		3	4
Ladies' club, ICA	1	2		3	4
Fund raising work	1	2		3	4
Active retirement group	1	2		3	4
Adult education group	1	2		3	4
Musical/drama/ dancing group	1	2		3	4
Political party	1	2		3	4
Other (specify)	1	2		3	4

Comment – Interviewer may add additional comments here.

Formal Education

In the context of looking at whether further education opportunities should be provided to older adults, I'd like to ask you about your own education.

7. a) What level of formal education did you reach when you were at school? [In the full-time formal education system]

Incomplete primary	1
Primary Cert or equivalent	2
Some 2nd level, no exams	3
Group/Inter Cert or Equiv	4
Leaving/Matric or Equiv	5
Diploma from Univ/3rd level Inst	6
Primary Degree	7
Higher Degree	8

b) Did you return to formal education after leaving school?

Yes … 1 No … 2

c) If 'yes' what was the level of study you did and at what age did you complete this?

Group/Inter/Junior level	1
Leaving Cert. level	2
Diploma level	3
Degree level	4
Higher degree level	5
Other [specify]	6 ______________________

d) Age of completion ______

8. a) Did you do any adult education classes since you left school?

Yes … 1 No … 2

If 'yes' Specify upto three main ones

1) ______________________________

2) ______________________________

3) ______________________________

b) Was there a course which was of particular benefit to you?

Yes … 1 Specify ______________________ No … 2

c) Would you enjoy the opportunity of further education courses?

Yes … 1 No … 2

If 'yes' what type of course might interest you [Specify]

____

Help given and help received

I'd like to ask you how long you have lived in your present home and who else lives with you.

9. a) How many years have you lived permanently in this locality? [1–2 miles] ________

b) What do you usually call the locality where you live? [e.g. Edenmore, Raheny, Coolock etc.] ________________

10. Have you any close relatives living in this locality? (1–2 miles). Approximately how many?
E.g. 1 brother, 2 sisters, 1 daughter etc. If none show 0

____

[Close relatives are parents, sons, daughters, sisters, brothers, grandchildren, aunts, uncles, nephews and nieces]

11. Do you live alone or with another/others?

Co-residence … 1 Alone … 2

If living alone move on to q. 14. If co-residing go on to q. 12.

12. Can I ask you about the other people who live in this household?

List by friend, relative or lodger, e.g. spouse, one daughter, two grandsons, cousin, lodger etc.

I'd like to ask you about the help you give and the help you receive with day-to-day activities.

13. a) Is there anyone in this household to whom you give care or mind, because of illness, handicap, old age or because they are children?

Yes … 1 No … 2

If 'yes' who is this? ______________________________

[***If more than one, record details***]

b) What is the nature of his/her illness/disability/dependency?

____

14. a) Do you provide any help or assistance with day-to-day activities to anybody not living in your household?

No … 2 Yes … 1

If 'yes' who is the principal person you help ____________________

b) How many people do you help on a regular basis….________

c) Thinking of the principal persons you help, which of the following things do you do?
(Can code any number)

Mind child/children	1
Personal care tasks (washing, dressing, help with toilet etc.)	2
Household chores	3
Provide a meal	4
Shopping	5
Other (errands post office, bank, library etc.)	6
Giving lift/bringing places	7
Looking after pet/bringing dog for walk	8
Calling to visit	9
Other (specify)	10

15. a) Do you have a home help?
[***Prompt – <u>if clear that person is active ask in a different way</u>***]

No … 2 Yes … 1

b) How many hours per week do they come in? _____ hours

c) Is this person paid? By whom?

Yes, I pay … 1 Yes, H.B. pay … 2 I and H.B. pay … 3
Yes, somebody pays on my behalf … 4 Not paid … 5

16. Is there anyone else who comes in regularly to give you care or assistance with day-to-day activities?

Yes … 1 No … 2

If 'yes' who is this? [relative, friend, neighbour etc.] ________________

General state of health and daily activities

I'd like to ask you now about your general state of health. Health is a factor that can have a big impact on people's quality of life. Other studies have shown that people can describe their general state of health in an accurate way.

17. How would you rate your present state of health? Would you say it is:

[Running Prompt]	Very good	1
	Good	2
	Fair	3
	Bad	4
	Very bad	5

18. a) Do you yourself have any major illness, physical disability or infirmity that has troubled you for at least the past year or that is likely to go on troubling you in the future?

No … 2 Yes … 1

If 'yes'
What is the nature of this illness/disability? ________________________

18. b) ***Interviewer: Is respondent:***

No mobility problem … 1 | Wheel chair user … 2

Other mobility problem … 3 | Confined to bed … 4

19. **Health problems may make it difficult for people to do some day-to-day things and this is a subject other studies have enquired into. So that we know whether the people who live in this area are similar to others who have been studied elsewhere, we want to ask you some questions which have been asked in these other studies.**

Could you tell me

a) the degree of difficulty if any you have in carrying out the following; and

b) who usually is your main helper if help is needed.

[Prompt – If the person has said they are in very good health and are active generally there is no need to go through entire list.]

Can do the following	Without difficulty	With difficulty	Only with help	Not at all	Main helper if Codes 2, 3, 4
Do grocery shopping	1	2	3	4	
Cook a hot meal	1	2	3	4	
Walk half a mile	1	2	3	4	
Get on a bus/train	1	2	3	4	
Get up and down steps	1	2	3	4	
Hear a conversation between several people (with hearing aid if necessary)	1	2	3	4	
Read a newspaper (with glasses if necessary)	1	2	3	4	
Dress self	1	2	3	4	
Have bath or shower	1	2	3	4	

20. a) If you have any problems or worries, do you have someone you can turn to that you can talk to about them?

Yes … 1 No … 2

If yes, who are the main [two] people? [relative, friend etc]

____

b) Are there any of your family, friends or relatives who often talk to you or rely on you for advice or support when they have personal problems or worries?

Yes … 1 No … 2

If yes, who are the main [two] people?

____

Social Activities and Contacts

I'd like to ask you about your interests and social activities.

21. a) Could you tell me if you participate in any of the following kinds of organisations or activities?
b) And if you do participate, about how many hours per week do you give to that participation?
[Interviewer – go through entire list and give details of the particular group/club in last column.]

Activity	No	Yes	Hours per Week Approx.	Details – name of group/activity/location
Outdoor sport (golf, tennis, pitch and put etc.)	2	1		
Walking, physical exercise	2	1		
Snooker / darts/ bowls or other indoor sports	2	1		
Active retirement group	2	1		

Senior citizen's club	2	1		
Ladies club/ICA	2	1		
Day centre	2	1		
Adult education/formal classes	2	1		
Going to the pub	2	1		
Scrabble/crosswords	2	1		
Music, dance, drama (Choir, drama group etc)	2	1		
Art/craft activity	2	1		
Bingo	2	1		
Card games (bridge etc.)	2	1		
Use of library	2	1		
Creative writing	2	1		
Gardening	2	1		
Reading books	2	1		
Reading newspapers	2	1		
Watching sport on television	2	1		
Watching television	2	1		
Listening to radio	2	1		
Going to cinema, theatre, concerts	2	1		
Computer	2	1		
DIY	2	1		
Playing music	2	1		
Listening to music	2	1		
Other activity-please specify	2	1		

21. c) Do you have any one or two major interests?

Yes … 1 Specify_______________________________________

No … 2

22. a) Do you drive?	Yes … 1	No … 2
b) Do you have a car or access to a car?	Yes … 1	No … 2

23. a) Are you a member of a particular church/religion?

No … 2 Yes … 1

If 'yes' which one ___________________________________

b) Do you attend mass or religious services and how often?

Daily	1
2–3 times per week	2
Once a week	3
Once or twice a month	4
A few times a year	5
Rarely/never	6
Listen to Mass on television/radio	7

Is religion or spirituality important to you?

Yes, very important	1
Fairly important	2
Not very important	3

Record any comment made

It is important to this study to build up a picture of people's networks of contacts with others. I'd like to ask you about your contacts with family members outside this household.

24. Could you tell me about any close relatives you have and where they live.

	Number living in this locality (under 3 miles)	Number living elsewhere in Dublin	Number living elsewhere in Ireland	Number living abroad
Daughters				
Sons				
Brothers or sisters				
Brothers-in-law or sisters-in-law				
Grandchildren				

Generally, I'd like to know how often you have contact with your children, other relatives, neighbours, friends and other people.

25. a) Do you have visits from:

	At least weekly	At least monthly	Few times a year	Rarely if ever	N/A
Daughters or sons	1	2	3	4	9
Other relatives	1	2	3	4	9
Friends	1	2	3	4	9
Neighbours	1	2	3	4	9
Clergy/religious	1	2	3	4	9
Voluntary group	1	2	3	4	9
Other (specify)	1	2	3	4	9

b) Do you make visits to the following:

	At least weekly	At least monthly	Few times a year	Rarely if ever	N/A
Daughters or sons	1	2	3	4	9
Other relatives	1	2	3	4	9
Friends	1	2	3	4	9
Neighbours	1	2	3	4	9
Other (specify)	1	2	3	4	9

26. Would a week go by without you either visiting someone else or having a visitor?

Yes … 1 No … 2

Record any comment made ______________________________

27. a) Do you have telephone contact with the following:

Note: Indicate if no telephone ______________________

	At least weekly	At least monthly	Few times a year	Rarely if ever	N/A
Daughters or sons	1	2	3	4	9
Other relatives	1	2	3	4	9
Friends	1	2	3	4	9
Neighbours	1	2	3	4	9
Other (specify)	1	2	3	4	9

b) Is there anyone you regularly write letters to? Yes … 1 No … 2

If 'yes' Specify ___________________________________

28. a) In this area, do people greet you when you go out in the street?

Yes … 1 No … 2

b) How important would it be for you to live in an area where you know people to see and they say hello to you in the street?

Very important … 1
Fairly important … 3
Not important … 2

29. a) What do you most like about living in this area?

[Specify] ______________________________

b) What do you most dislike about living in this area?

[Specify] ______________________________

30. a) Which of the following statements best describes this locality, taking the locality as the next few streets around your house:
[Show on separate sheet – respondent to select one only]

People are both friendly and helpful to each other in this locality	1
People keep to themselves in this locality	2
People don't really know each other in this locality	3
There was more friendliness and helpfulness in this locality in the past than there is now	4
It is difficult to lead your own life in this locality because people try to intrude too much	5
People are friendly in general but they let you lead your own life	6
There is little friendliness or neighbourliness in this locality	7

b) Do you have a sense of safety or security in this area?
Yes … 1 No … 2 To some degree/not completely … 3

c) Do you have a sense of safety or security in your home?
Yes … 1 No … 2 To some degree/not completely … 3

General experience of ageing

I would like to ask you about your general experiences of growing older in terms of how you feel and how other people treat you. Again, these are questions which have also been asked in other surveys, which will allow us to compare results.

31. Firstly, could you tell me whether you agree or disagree with the following statements: [***call out each statement***]

Statement	Agree	Disagree	Don't know
a) Older people are treated respectfully in Ireland in my experience	1	2	3
b) Retirement has given me a new lease of life	1	2	3
c) Old age has given me a new lease of life	1	2	3
d) Younger people are helpful towards older people	1	2	3
e) Older people should stand up more actively for their rights	1	2	3
f) Families are less willing to care for older relatives than they used to be	1	2	3

32. I would like to ask you whether you have enough to fill your day or whether you have too much time on your hands. Which of the following statements best describes your day: ***[Show on separate sheet – respondent to select one only]:***

Statement	Code
There is hardly enough time in the day. I'm very busy.	1
I have a full day but not too much to do.	2
I have plenty of free time.	3
I have so little to do. I often have too much time on my hands.	4
I have nothing to do	5
Don't know	6

33. Loneliness is a problem for people at many different stages of life, and for many different reasons. Other surveys have explored whether later life is connected with problems of this kind or not.

Do you experience loneliness?

Frequently … 1
Sometimes … 2
Rarely/never … 3

Record any comment made ________________________________

34. a) Have you ever considered taking on part-time work, paid or unpaid?

Yes … 1 No … 2 Maybe … 3

Record any comment made ________________________________

b) Are there events during the week or which happen fairly often which you do for enjoyment or fun?
Could you tell me a bit about them?

Record any comment made ________________________________

c) Can you tell us about any plans for the future that you are looking forward to?

Record any comment made ________________________________

Use of health and social services

I'd like to ask you about your use of and experience of various health and social services:
[*Note: Interviewer to use discretion in relation to how this question is asked*]

35. Do you use any of the following services?
a) How often? And b) what has been your general level of satisfaction with them?
Code for each service a) Frequency b) General satisfaction

Service	Regular use	Occa-sional use	Satis-fied	Fairly satis-fied	Not satis-fied	N/A
Public health or district liaison nurse	1	2	3	4	5	6
GP	1	2	3	4	5	6
Physiotherapy	1	2	3	4	5	6
Home help	1	2	3	4	5	6
Hospital	1	2	3	4	5	6
Day care centre	1	2	3	4	5	6
Day centre	1	2	3	4	5	6
Transport to/from hospital/ other soc. service	1	2	3	4	5	6
Occupational therapy	1	2	3	4	5	6
Chiropody	1	2	3	4	5	6
Meals service	1	2	3	4	5	6
Social worker	1	2	3	4	5	6
Social welfare office/ pensions and other entitlements	1	2	3	4	5	6
Warden/Local authority	1	2	3	4	4	6
Other – specify						

Future accommodation/care needs

I'd like to ask you about your future accommodation needs and your preferences.

36. a) In your present circumstances, given the opportunity would you like to move house or otherwise change your living arrangements?

Yes, definitely … 1   Maybe/it depends … 2   No … 3

b) If 'yes' or 'maybe' what type of move or change would you like to make?

To bigger house/flat	1
To smaller house/flat	2
To a different area	3
To residential home	4
To sheltered housing	5
To accommodation better adapted for disability	6
Move in with relatives	7
Have relatives move in here	8

37. If you were no longer able to care for yourself what would be your preferred living arrangement:
Rank any number in order of preference – 1st choice, 2nd etc.

a) To residential nursing home ____________
b) To sheltered housing ____________
c) Move in with relatives ____________
d) Have relatives move in here ____________
e) Continue to live at home with care support ____________
f) Other (specify) ____________

38. Would you say that you enjoy your life from day to day?

Very much	1
Quite a lot	2
Sometimes	3
Rarely/never	4

Record any comment made

____

39. a) Have you any particular ideas about how older people could be more involved in their community?

____

____

Are there important aspects to being an older person living in this area that we haven't covered in this interview?

____

____

40. Finally, can I ask you whether you would be prepared to be interviewed further in connection with this study.

Yes … 1 Maybe … 2 No … 3

If 'yes' or 'maybe' would you be happy to give me your name and phone number? These will not be given to any person outside this project.

____

Part B – Overall Impressions

Are there any aspects of what this interviewee said which you found especially interesting or which might be followed up later?
Please record any aspects of the interview with this particular person which you found striking.

____

____

____

____

____

____

Part C – Details of Accommodation and General Environment

Part C to be completed by interviewer based on observation

1. Who was present during the interview?

Sampled person only … 1 Sampled person plus other(s) … 2
Specify other(s) ______________________________

2. Is there anybody else over 65 years living in the household or a neighbouring house who could be approached to answer a questionnaire?
Yes … 1 No … 2 Specify ____________________

3. What type of building is respondent's household located in?

Detached house/bungalow	1
Semi-detached house/bungalow	2
Terrace house	3
Block of purpose built flats	4
Other (specify)	5

4. Is house/apartment
Local authority built … 1 Privately built … 2

5. Appearance of immediate neighbourhood, including other houses, gardens and open areas:

Very well maintained, tidy houses and gardens and open areas with trees, shrubs etc.	1
Reasonably well maintained, tidy houses and gardens and some open areas with trees, shrubs etc.	2
Generally poorly maintained houses and gardens with very few trees, shrubs etc.	3

Part D – Profile of Respondent

Age of Respondent

Sex (M or F)

Living alone/co-residence (Alone or Co)

Appendix B: Organisations Representing Older People

Age Action Ireland:
Involved in campaigning, research, Third Age learning, carer support and information services.

Age and Opportunity:
Promotes positive attitudes to ageing and greater participation by older people in society. Promotes and coordinates programmes for sport, physical activity and the arts.

Carers Associations:
Provide advice and support for carers (not exclusively involving older people) and lobby for more state supports for carers.

Federation of Active Retirement Associations:
Promotes and coordinates Active Retirement groups. Its aims are to help older people live happy and healthy retirement through a wide range of activities and community work. Name recently changed to Active Retirement Ireland.

Irish Association of Older People:
A membership based campaigning group for older people. It represents the interests of all older people but especially those who are less articulate and more vulnerable.

Irish Senior Citizens' Parliament:
National campaigning body representing older people's issues to Government. It has over 200 affiliated organisations, principally Active Retirement groups, Senior Citizens' groups and Pensioners' associations.

National Council on Ageing and Older People:
Advisory agency to the government on all aspects of ageing and the welfare of older people. Also undertakes research.

Senior Help Line:
Provides a confidential and friendly listening service to older people. It is manned by older people themselves and is based in Summerhill, Co. Meath. Has recently expanded to other parts of the country.

Appendix C: Frequency tables

Table A8.1: Frequency table for five items on index

No. of items	Frequency	Per cent
.00	2	1.2
1.00	25	15.2
2.00	42	25.5
3.00	36	21.8
4.00	43	26.1
5.00	17	10.3
N = 165		

Table A8.2: Frequency table for four items on index

No. of items	Frequency	Per cent
.00	16	9.7
1.00	47	28.5
2.00	39	23.6
3.00	46	27.9
4.00	17	10.3
N = 165		

Appendix D: Note on statistical analysis used

The multivariate parametric statistical analysis involved reliability analysis and K Means Cluster analysis. The reliability test on the five variables suggested that the five variables do act cohesively. However, when we omitted 'visiting contact' from the analysis and carried out a reliability test for four items the Alphas for the four or five items together were equally good (Alpha = .4851 for five items and Alpha = .4662 for four items) suggesting that either four or five items could be put together to form a total index.

We used latent class cluster analysis to see how the variables selected to form the index of connectedness would differentiate and group our respondents. The statistical analysis showed that two models fit the survey data; when five variables were used 3 clusters emerged and when four variables were used 2 clusters emerged.

Bibliography

Age and Opportunity, 2003, *Home from Home? The Views of Residents on Social Gain and Quality of Life: A Study in Three Care Centres for Older People*, Dublin: Age and Opportunity.

Ageing in Ireland, 2007, Ireland: Central Statistics Office.

Allan, G., 1996, *Kinship and Friendship in Modern Britain*, Oxford: Oxford University Press.

—— 1998, 'Friendship and the private sphere', in Adams, R. and Allan, G. (eds), *Placing Friendship in Context*, UK: Cambridge University Press.

Arber, S., Davidson, K. and Ginn, J., 2003, 'Changing Approaches to Gender and Later Life', in Arber, S., Davidson, K. and Ginn, J. (eds), *Gender and Ageing – Changing Roles and Relationships*, UK: Open University Press.

—— and Attias-Donfut, C., 2000 (eds), *The Myth of Generational Conflict – the family and state in ageing societies*, UK: Routledge/ESA Studies in European Societies.

—— and Evandrou, M, 1993, 'Mapping the Territory – Ageing, Independence and the Life Course', in Arber, S. and Evandrou, M. (eds), *Ageing, Independence and the Life Course*, UK: Jessica Kingsley in association with the British Society of Gerontology.

Arensberg, C. and Kimball, S., 1968, *Family and Community in Ireland*, Ennis, Ireland: Clasp Press.

Bales, S., Eklund, S. J. and Siffin, C. F., 2000, 'Children's perceptions of elders before and after schools-based intergenerational program'. *Educational Gerontology*, Vol. 26, pp. 677–689.

Ballymun Active Living Survey, Ballymun Regeneration Ltd., 1999.

Bellah. R., Madsen, R., Sullivan, W., Swidler, A. and Tipton, S., 1996 [1985], *Habits of the Heart: Individualism and Commitment in American Life*, Berkeley: University of California Press.

Bertaux, D., 1982, 'The Life Course Approach as a Challenge to the Social Sciences', in Hareven, T. K. and Adams, K. (eds), *Ageing*

and Life Course Transitions – An Interdisciplinary Perspective, London: Tavistock.
Bond, J., Coleman, P. and Peace, S. (eds), 1993, *Ageing in Society: An Introduction to Social Gerontology*, Second edition, London: Sage.
Bould, S., 2003, 'Caring Neighbourhoods: Bringing up the Kids Together', *Journal of Family Issues*, Vol. 24, no. 4, pp. 427–47.
Bowling, A., Gabriel, Z., Dykes, L. M., Evans, O., Fleissig, A., Banister, D. and Sutton, S., 2003, 'Let's ask them: a national survey of definitions of quality of life and its enhancement among people aged 65 and over', *International Journal of Ageing and Human Development*, Vol. 56, no. 4, pp. 269–306.
Brown, C., 2001, *The Death of Christian Britain*, London: Routledge.
Bulmer, Martin, 1986, *Neighbours: the work of Philip Abrams*, Cambridge: Cambridge University Press.
Burke, H., 1999, 'Foundation Stones of Irish Social Policy, 1831–1951', in G. Kiely, A. O'Donnell, P. Kennedy and S. Quin (eds), *Irish Social Policy in Context*, Dublin: University College Dublin Press.
Bygren, L. O., Konlaan, B. B. and Johansson S. E., 1996, 'Attendance at cultural events, reading books or periodicals, and making music or singing in a choir as determinants for survival: Swedish interview survey of living conditions', *British Medical Journal*, Vol. 313, pp. 1577–1580.
Byrne, A., Edmondson, R. and Varley, T., 2001, Introduction to the third edition of Arensberg, C. and Kimball, S. (1968) *Family and Community in Ireland*, Ennis, Ireland: Clasp Press.
Calhoun, C. J., 1980, 'Community: toward a variable conceptualisation for comparative research', *Social History*, Vol. 5, pp. 105–129.
Care of the Aged, 1968, Report of Interdepartmental Committee, Dublin: Stationery Office.
Central Statistics Office, 2002, Census of Population reports, Ireland.
Cicero, *De Senectute* (translated by W. A. Falconer), 1923, Loeb classical library.
Clarke, L. and Cairns, H., 2001, 'Grandparents and the Care of Children: the Research Evidence', in Broad, B. (ed), *Kinship Care –*

The placement choice for children and young people, London: Russell House Publishing.

Coleman, P. G., 2006, *Spirituality, Health and Ageing*, David Hobman Memorial Lecture, Age Concern England and King's College London.

—— Mc Kiernan, F., and Speck, P., 2002, 'Spiritual belief and quality of life: the experience of older bereaved spouses', *Quality in Ageing – Policy, Practice and Research*, Vol. 3, no. 1, pp. 20–26.

—— Ivani-Chalian, C. and Robinson, M., 2004, 'Religious attitude among British older people: stability and change in a 20-year longitudinal study', *Aging and Society*, Vol. 24, pp. 167–188.

Collins, C. and Shelley, E., 1997, 'Social Class Differences in Lifestyle and Health Characteristics', in Tracey, M. and Cleary, A. (eds), *The Sociology of Health and Illness in Ireland*, Dublin: University College Dublin Press.

Connell, P. and Pringle, D., 2004, *Population Ageing in Ireland: Projections 2002–2021*, Dublin: National Council on Ageing and Older People.

Conroy, P., 1999, 'From the 50s to the 90s – Social Policy Comes out from the Shadows', in Kiely, G., O'Donnell, A., Kennedy, P. and Quin, S. (eds), *Irish Social Policy in Context*, Dublin: University College Dublin Press.

Convery, J., 1987, *Choices in Community Care – Day Centres for the Elderly in the Eastern Health Board*, Dublin: National Council for the Aged, Report no. 17.

Corcoran, M., Gray, J. and Peillon, M., 2007, 'Ties that Bind? The Social Fabric of Daily Life in New Suburbs', in Fahey, T., Russell, H. and Whelan, C. (eds), *Best of Times? The Social Impact of the Celtic Tiger*, Dublin: Institute of Public Administration.

Crow, G., 2004, 'Social Networks and Social Exclusion: An Overview of the Debate', in Phillipson, C., Allan, G. and Morgan, D., *Social Networks and Social Exclusion-Sociological and Policy Perspectives*, UK: Ashgate.

Cumming, E. and Henry, W., 1961, *Growing Old: the Process of Disengagement*, New York: Basic Books.

Curtin, C. and Ryan, C., 1987, 'Clubs, Pubs and Private Houses in a Clare Town', in Curtin, C. and Wilson, T. (eds), 1987, *Ireland From Below – Social Change and Local Communities*, Galway: Galway University Press.

Dant, T., 1988, 'Dependency and old age: theoretical accounts and practical understandings'. *Ageing and Society*, Vol. 8, pp. 171–188.

Davey-Smith, G., Blane, D. and Bartley, M., 1994, 'Explanations for Socio-Economic Differentials in Mortality: Evidence from Britain and Elsewhere', *European Journal of Public Health*, Vol. 4, pp. 132–144.

Davidson, K., Daly, T. and Arber, S., 2003, 'Older Men, Social Integration and Organisational Activities', *Social Policy and Society*, Vol. 2, no. 2, pp. 81–89.

Davie, G., 2000, *Religion in Modern Europe*, Oxford: OUP.

Department of Health, 1988, Health (Nursing Homes) Act 1990, Dublin: Stationery Office.

—— 1988, *The Years Ahead: A Policy for the Elderly*, Dublin: Stationery Office.

—— 1994, *Shaping a Healthier Future: A Strategy for Effective Health Care in the 1990s*, Dublin: Stationery Office.

Department of Health and Children, 1998, *Adding Years to Life and Life to Years – Health Promotion Strategy for Older People*, Dublin: Stationery Office.

—— 2001, *Quality and Fairness – A Health System For You*, Dublin: Stationery Office.

—— 2007, Draft National Standards for Residential Care Settings for Older People, Dublin: Stationery Office.

Department of Social, Community and Family Affairs, 2000, Supporting Voluntary Activity – A White Paper on a Framework for Supporting Voluntary Activity and for Developing the Relationship between the State and the Community and Voluntary Sector, Dublin: Stationery Office.

Dillon, M. and Wink, P., 2007, *In the Course of a Lifetime-Tracing Religious Belief, Practice and Change*, Berkeley: University of California Press.

Eastern Health Board, 1999, *Ten Year Action Plan for Services for Older Persons 1999–2008*, Dublin: EHB.
Eastern Health Board, 2001, *Review of the Implementation of the Ten Year Action Plan for Services for Older People*, Eastern Regional Health Authority, Dublin: ERHA.
Edmondson, R., 2000, 'Silence and the Golden Years: Older People and Public Policy in Ireland', in Taylor, G. (ed), *Issues in Irish Public Policy*, Dublin: Irish Academic Press.
—— 2001, 'Studying Civic Culture Ethnographically and What it Tells Us about Social Capital' in Dekker, P. and Uslaner, E. M. (eds), *Social Capital and Participation on Everyday Life*, London and New York: Routledge.
Edwards, B. and Foley, M. W., 1997, 'Social Capital and the Political Economy of Our Discontent', *American Behavioral Scientist*, Vol. 40, no. 5.
Elias, N., 1974, 'Foreword –Towards a Theory of Communities', in Bell, C. and Newby, H. (eds), *The Sociology of Community: A Selection of Readings*, London: Frank Cass.
—— and Scotson, J., 1965, *The Established and the Outsiders*, London: Frank Cass.
Erikson, E. H., 1980, *Identity and the Life Cycle: a Reissue*, New York: W.W. Norton.
—— Erikson, J. M. and Kivnick, H. Q., 1986, *Vital Involvement in Old Age*, New York: W.W. Norton.
Estes, C., 1979, *The Ageing Enterprise*, Jossey-Bass: San Francisco.
—— Biggs, S. and Phillipson, C., 2003, *Social Theory, Social Policy and Ageing – a critical introduction*, England: Open University Press.
Eurobarometer, 1993, *Age and Attitudes: Main Results of a Eurobarometer Survey*, Brussels: Directorate-General 5, European Commission.
Evason, E. and G. Robinson, 1996, 'Informal Care in Northern Ireland'. In Breen, R., Devine, P. and O'Dowd L. (eds), *Social Attitudes in Northern Ireland*, Belfast: Appletree Press.
Fahey, T., 2002, 'Is atheism increasing? Ireland and Europe compared' in Cassidy, E. (ed), *Measuring Ireland: Discerning values and beliefs*, Dublin: Veritas.

—— and Murray, P., 1994, *Health and Autonomy Among the Over 65s in Ireland*, Dublin: National Council for the Elderly, Report no. 39.
—— and Russell, H., 2001, *Older People's Preferences for Employment and Retirement in Ireland*, Dublin: Economic and Social Research Institute, Report no. 67.
—— and Fitzgerald, J., 1997, *Welfare Implications of Demographic Trends*, Dublin: Oak Tree Press/Combat Poverty Agency.
Fanning, B., 1999, 'The Mixed Economy of Welfare', in Kiely, G., O'Donnell, A., Kennedy, P. and Quin, S. (eds), *Irish Social Policy in Context*, Dublin: University College Dublin Press.
Faughnan, P. and Kelleher, P., 1993, *The Voluntary Sector and the State: a study of organisations in one region*, Dublin: Conference of Major Religious Superiors.
Featherstone, M. and Hepworth, M., 1989, 'Ageing and Old Age: Reflections on the Postmodern Life Course', in Bytheway, B., Keil, T., Allatt, P. and Bryman, A. (eds), *Becoming and Being Old – Sociological Approaches to Later Life*, London: Sage.
Feldman, S., Mahoney, H. and Seedman, T., 2003, 'School-based Intergeneration programs: laying the foundations for sustainability'. *Journal of Intergenerational Relationships*, Vol. 1, no. 2.
Field, Schuller, Baron, 2000, 'Social Capital and Human Capital Revisited', in Baron, S., Field, J. and Schuller, T. (2000), *Social Capital – Critical Perspectives*, Oxford: Oxford University Press.
Finch, J., 1989, *Family Obligations and Social Change*, UK: Polity Press in association with Basil Blackwell.
Fischer, C. S., 1982, *To Dwell Among Friends*, Chicago: University of Chicago Press.
Friedman, M., 1992, 'Feminism and Modern Friendship: Dislocating the Community', in Avineri, S. and de-Shalit, A. (eds), *Communitarianism and Individualism*, Oxford: Oxford University Press.
Gallagher, C. and Kennedy, K., 2003, 'The training implications of a social care approach to working with older people', *Irish Journal of Applied Social Studies*, Vol. 4, no. 1, pp. 21–35.
Gallagher, C., 2005, 'Social Care Work and the Older Person', in Share, P. and McElwee, N. (eds), *Applied Social Care – An Introduction for Irish Students*, Dublin: Gill and Macmillan.

—— 2006, 'Social Policy and a Good Life in Old Age', in O'Dell, E. (ed), *Older People in Modern Ireland. Essays on Law and Policy*, Dublin: First Law.

Gans, H., 1962, *The Urban Villagers*, Glencoe, Illinois: The Free Press.

Garavan, R., Winder, R. and McGee, H., 2001. *Health and Social Services for Older People (HeSSOP)*, Dublin: National Council on Ageing and Older People.

Gauntlett. D. and Hill A., 1999, *TV Living – Television, Culture and Everyday Life*, London: Routledge.

Gill, 1999, *Churchgoing and Christian Ethics*, Cambridge: Cambridge University Press.

Gilleard, C. and Higgs, P., 2005, *Contexts of Ageing – Class, Cohort and Community*, UK: Polity Press.

Granovetter, M. S., 1973, 'The Strength of Weak Ties', *American Journal of Sociology*, Vol. 78, no.6, pp. 1360–1380.

Grundy, E., 1995, Demographic Influences on the Future of Family Care, in I. Allen and E. Perkins, *The Future of Family Care of Older People*, London: HMSO.

Guinnane, T. W., 1996, 'The Family, State Support and Generational Relations in Rural Ireland at the Turn of the Twentieth Century', in Hareven, T. K. (ed), *Aging and Generational Relations. Life-Course and Cross-Cultural Perspective*, Berlin and New York: Walter de Gruyter and Co.

Hall, P., 1999, 'Social Capital in Britain', *British Journal of Political Science*, Vol. 29, pp. 417–461.

Hannan, D., 1972, 'Kinship, neighbourhood and social change in Irish rural communities', *Economic and Social Review*, Vol. 3, pp. 163–187.

Hareven, T. K. (ed), 1996, *Aging and Generational Relations. Life-Course and Cross-Cultural Perspective*, Berlin and New York: Walter de Gruyter and Co.

—— and Adams, K., 1996, 'The Generation in the Middle: Cohort Comparisons in Assistance to Ageing Parents in an American Community', in Hareven, T. K. (ed), 1996, *Aging and Generational Relations. Life-Course and Cross-Cultural Perspective*, Berlin and New York: Walter de Gruyter and Co.

Haslett, D., 2003, *The Role and Future Development of Day Services for Older People in Ireland*, Dublin: National Council on Ageing and Older People, Report no. 74.

Havighurst, R. J., 1963, 'Successful Ageing', in Williams, R. H., Tibbits, C. and Donohue W. (eds), *Processes of Aging*, Vol. 1, pp. 299–320, New York: Atherton,.

Herbert, D., 2003, *Religion and Civil Society – Rethinking Public Religion in the Contemporary World*, London: Ashgate.

Inglis, T., 1998, *Moral Monopoly – The Rise and Fall of the Catholic Church in Modern Ireland*, Dublin: Gill and Macmillan.

Institute of European Food Studies, 1999, *Consumer Attitudes to Physical Activity, Body-Weight and Health*, Institute of European Food Studies.

Iso, A., Jackson, E. and Dunn, E., 1994, 'Starting, Ceasing and Replacing Leisure Activities over the Life-Course', *Journal of Leisure Research*, Vol. 26, no. 3, pp. 227–249.

Jerrome, D., 1992, *Good Company: an anthropological study of older people in groups*. Edinburgh: Edinburgh University Press.

Johnson, P., 1987, *Structured Dependency of the Elderly: a Critical Note*, Centre for Economic Policy Research, London.

Kaplan, M., 1979, *Leisure: Lifestyle and Lifespan*, Philadelphia: W. B. Saunders.

Keating, N., Fast, J., Forbes, D. and Wenger, C., 2002, *Informal Care Networks of Canadian Seniors with Long-Term Health Problems*, Final Report submitted to the National Health Research and Development Programme, Canada.

Lash, S. and Urry, J., 1987, *The End of Organised Capitalism*, Cambridge: Polity.

Laslett, P., 1989, *A Fresh Map of Life*, UK: George Weidenfeld and Nicolson Ltd.

—— 1996, *A Fresh Map of Life – The Emergence of the Third Age*, UK: Macmillan Press Ltd.

Layte, R., Fahey, T. and Whelan, C., 1999, *Income, Deprivation and Well-being Among Older Irish People*, National Council on Ageing and Older People, Report no. 55, Dublin.

Lee, D. and Newby, H., 1983, *The Problem of Sociology*, London and New York: Routledge.

Limerick County Development Board, Survey of Attitudes of County Limerick Residents, Limerick County Development Board, 2001.

Lubarsky, N., 1997, Rememberers and remembrances: Fostering connections with intergenerational interviewing, in Brabazon K. and Disch. R. (eds), 1997. *Intergenerational Approaches in Ageing: Implications for Education, Policy and Practice*, New York: Haworth Press.

Lundstrom, F., 1995, *Home Help Services for Elderly People in Ireland – Proceedings of Conference*, National Council for the Elderly, Dublin, Report no. 41.

Lundstrom, F., 2001, *Grandparenthood in Modern Ireland*, Dublin: Age Action Ireland.

—— and McKeown, K., 1994, *Home Help Services for Elderly People in Ireland*, Dublin: National Council on Ageing and Older People.

McAleese, Mary, 2005, Address by President Mary McAleese at the General Assembly of the Conference of Religious in Ireland in the Great Southern Hotel, Rosslare Strand, Co. Wexford, 6th June, 2005.

Mac Éinrí, P., 2006, 'Migration in Ireland: A Changing Reality', in Healy, S., Reynolds B. and Collins, M. (eds), *Social Policy in Ireland – Principles, Practice and Problems*, Dublin: The Liffey Press

MacIntyre, A., 1981, *After Virtue: A Study of Moral Theory*, London: Duckworth.

Mobilate Survey 2000 – *Enhancing Outdoor Mobility in Later Life-Health and Leisure Activities of Persons 55+ in Finland, Germany, Hungary, Italy and the Netherlands*, European Commission.

Moody, H. R., 1988, *Abundance of Life*, New York: Columbia University Press.

—— 1998, *Ageing Concepts and Controversies*, Second edition, USA: Pine Forge Press.

Murphy, C., O'Shea, E., Cooney, A., Shiel, A. and Hodgins, M., 2006, *Improving quality of life for older people in long stay care settings in Ireland*, Dublin: National Council on Ageing and Older People, Report no. 93.

National Council for the Elderly, 1985, *Institutional Care of the Elderly in Ireland*, Dublin: National Council for the Elderly.
National Council on Ageing and Older People, 1999, *The Future Organisation of Home Help Services in Ireland*, Report no. 53. Dublin: National Council on Ageing and Older People.
National Economic and Social Forum, 2003, *The Policy Implications of Social Capital*, Forum Report no. 28, Dublin: Stationery Office.
Nisbet, R., 1969, *The Quest for Community*, US: Oxford University Press.
O'Connor, P., 1992. *Friendship between women: a critical review*, New York: Harvester Wheatsheaf.
—— 1998, 'Women's friendships in a post-modern world', in Adams, R. and Allan, G. (eds), *Placing Friendship in Context*, UK: Cambridge University Press.
O'Shea, E., 2003, *Healthy Ageing in Ireland: Policy, Practice and Evaluation*, Dublin: National Council on Ageing and Older People, Report no. 77.
O'Shea, E., 2002, *Review of the Nursing Home Subvention Scheme*, Dublin: Stationery Office.
O'Shea, E., 1993, *The Impact of Social and Economic Policies on Older People in Ireland.* Dublin: National Council for the Elderly.
O'Shea, E., 2006, 'Public Policy for Dependent Older People in Ireland', in O'Dell, E. (ed), *Older People in Modern Ireland. Essays on Law and Policy.* Dublin: First Law.
O'Sullivan, L., 2003, *The Best Years of their Lives? Senior Citizens – Their Future and Their Past*, Dublin: St Andrews Resource Centre.
Pahl, R., 1998, 'Friendship: the Social Glue of Contemporary Society', in Franklin, J., *The Politics of Risk Society*, Cambridge: Polity Press.
—— 2000, *On Friendship*, UK: Polity.
—— and Spencer, L., 1997, 'The Politics of Friendship', *Renewal*, Vol. 5, no. 3, pp. 100–107.
Pargament, K. I., 1997, *The Psychology of Religion and Coping*, Guilford: New York.

Phillipson, C., 1982, *Capitalism and the Construction of Old Age*, London: Macmillan.

—— 1998, *Reconstructing Old Age – New Agendas in Social Theory and Practice*, London: Sage Publications.

—— Bernard, M., Phillips, J. and Ogg, J., 2001, *The Family and Community Life of Older People*, London and New York: Routledge.

—— 2004, 'Social Networks and Social Support in Later Life', in Phillipson, C., Allan G. and Morgan, D. (eds), *Social Networks and Social Exclusion-Sociological and Policy Perspectives*, UK: Ashgate.

Portes, Alejandro, 1998, 'Social Capital: Its Origins and Applications in Modern Sociology', *Annual Review of Sociology*, 1998, Vol. 24, pp. 1–24.

Powell, F. and Guerin, D., 1997, *Civil Society and Social Policy*, Dublin: A & A Farmar.

Protestants in Community Life – Findings from a Co. Donegal Survey, 2001, Derry and Raphoe Action: Raphoe, Co. Donegal.

Putnam, R., 2000. *Bowling Alone: The Collapse and Revival of American Community*, New York: Simon and Schuster.

Ruddle, H., Donoghue, F., and Mulvihill, R., 1997. *The Years Ahead Report: a review of the implementation of its recommendations*, Dublin: National Council on Ageing and Older People.

Scharf, T., Phillipson, C. and Smith, A., 2004, 'Poverty and social exclusion – growing older in deprived urban neighbourhoods', in Walker, A. and Hagan Hennessy, C. (eds), *Growing Older – Quality of life in old age*, UK: Open University Press.

Schuller, T. and Young, M., 1991, *Life after Work: the Arrival of the Ageless Society*, London: Harper-Collins.

Share, P., Tovey, H. and Corcoran, M., 2007, *A Sociology of Ireland*, Dublin: Gill and Macmillan.

Sundstrom, G., 1994. 'Care by Families: An Overview of Trends'. In *Caring for Frail Elderly People, New Directions in Care*, Paris. OECD.

Swisher, R., Sweet, S. and Moen, P., 2004, 'The Family-friendly community and its life-course fit for dual-earner couples', *Journal of Marriage and the Family*, Vol. 66, pp. 281–92.

Taylor, C., 1994, *Sources of the Self – The Making of the Modern Identity*, New York: University of Cambridge.

Tipping the Balance – Report and Recommendations to Government on Supporting and Developing Volunteering in Ireland, October, 2002, National Committee on Volunteering in Ireland: Dublin.

Tonnies, F., 1967, *Community and Society*, Michigan: Michigan State University Press, reprinted in Bell, C. and Newby, H. (eds), 1974, *The Sociology of Community: A Selection of Readings*, London: Frank Cass.

Townsend, P., 1981, 'The Structured Dependency of the Elderly: Creation of Social Policy in the Twentieth Century', *Ageing and Society*, Vol. 1, pp. 5–28.

—— 1986, 'Ageism and social policy'. In Phillipson, C. and Walker A. (eds), *Ageing and Social Policy*, Aldershot: Gower Press.

Treacy, P., Butler, M., Byrne, A., Drennan, J., Fealy, G., Frazer, K. and Irving, K., 2004, *Loneliness and Social Isolation Among Older Irish People*, Dublin: National Council on Ageing and Older People, Report no. 84.

Victor, C. R., Bowling, C. R. and Bond, J., 2002, *Loneliness, Social Isolation and Living Alone in Later Life*, Sheffield: ESRC.

Voster, J. and O'Searcaigh, C., 2003, *Caiseal na gCorr – Photographs by Jan Voster and Writings by Cathal O Searcaigh*, Cló Lar-Chonnachta, grant assisted by Donegal County Council, the Arts Council and Udaras na Gaeltachta.

Walker, A., 1981, 'Towards a political economy of old age'. *Ageing and Society*, Vol. 1, pp. 73–94.

—— 1993, 'Poverty and Inequality in Old Age', in Bond, J., Coleman, P., and Peace, S. (eds), *Ageing in Society: An Introduction to Social Gerontology*, Second edition, London: Sage.

Weiss, R., 1973, *Loneliness: The Experience of Social and Emotional Loneliness*, USA: MIT Press.

Wellman, B., 1990, 'The place of kinfolk in personal community settings', in Wellman, B. (ed), *Families in Community Settings: Interdisciplinary Settings*, New York: Haworth Press

Wenger, G. C., 1989, 'Support Networks in Old Age-constructing a typology', in Jefferys, M. (ed), *Growing Old in the 20th Century*, London: Routledge.

—— 1995, 'A Comparison of Urban with Rural Networks in North Wales', *Ageing and Society*, Vol. 15, pp. 59–82.

Whelan, C. T. and Whelan, B. J., 1988, *The Transition to Retirement*, Dublin: The Economic and Social Research Institute, Report no. 138.

Wilmott, P., 1987, *Friendships, Networks and Social Support*, London; Policy Studies Institute.

World Health Organisation, 2002, *Active Ageing: A Policy Framework* (http://www.who.int/hpr/ageing/publications.htm).

Wright, P., 1982, 'Men's Friendships, Women's Friendships and the Alleged Inferiority of the Latter', *Sex Roles*, Vol. 8, no.1, 1–20.

Index